PAINTING
OF THE
HIGH RENAISSANCE
IN
ROME AND FLORENCE

VOLUME II

PAINTING OF THE HIGH RENAISSANCE IN ROME AND FLORENCE

S. J. FREEDBERG

Icon Editions
Harper & Row, Publishers
New York, Evanston, San Francisco, London

This book was first published by Harvard University Press.

First ICON EDITION published 1972.

PAINTING OF THE HIGH RENAISSANCE IN ROME AND FLORENCE. Volume II. Copyright © 1961 by the President and Fellows of Harvard College. All rights reserved. Printed in the United States of America. No part of this book may be used or reproduced in any manner whatsoever without written permission except in the case of brief quotations embodied in critical articles and reviews. For information address Harper & Row, Publishers, Inc., 49 East 33rd Street, New York, N.Y. 10016. Published simultaneously in Canada by Fitzhenry & Whiteside Limited, Toronto.

STANDARD BOOK NUMBER: 06-430014-5

ILLUSTRATIONS

I

INTRODUCTION
THE GENESIS OF HIGH RENAISSANCE CLASSICAL STYLE

1. LEONARDO *Head of an Angel*, from Verrocchio's *Baptism of Christ*, Florence, Uffizi. Photo Sopr. alle Gallerie, Florence
2. VERROCCHIO *Baptism of Christ*, Florence, Uffizi. Photo Sopr. alle Gallerie, Florence
3. LEONARDO *Adoration of the Magi*, Florence, Uffizi. Photo Sopr. alle Gallerie, Florence
4. LEONARDO *Adoration of the Magi* (detail). Photo Alinari
5. LEONARDO *Adoration of the Magi* (detail). Photo Alinari
6. DOMENICO GHIRLANDAIO *Adoration of the Magi*, Florence, Innocenti. Photo Alinari
7. ANTONIO POLLAIUOLO *Hercules and the Hydra, Hercules and Anteus* (lost, formerly Florence, Uffizi). Photo Alinari
8. PIERO DELLA FRANCESCA *Montefeltro Altar*, Milan, Brera. Photo Alinari
9. BOTTICELLI *Adoration of the Magi*, Florence, Uffizi. Photo Sopr. alle Gallerie, Florence
10. LEONARDO *Virgin of the Rocks*, Paris, Louvre. Photo Alinari
11. LEONARDO *Virgin of the Rocks* (detail). Photo Alinari
12. LEONARDO *Last Supper*, Milan, S.M. delle Grazie. Photo Alinari
13. LEONARDO *Last Supper* (detail). Photo Alinari
14. FILIPPINO LIPPI *Adoration of the Magi*, Florence, Uffizi. Photo Brogi
15. PIERO DI COSIMO *Mars and Venus*, Berlin, Museums. Photo museum
16. PERUGINO *Vision of St. Bernard*, Munich, Pinakothek. Photo museum
17. MICHELANGELO *Madonna of the Steps*, Florence, Casa Buonarroti. Photo Sopr. alle Gallerie, Florence
18. MICHELANGELO *Battle of the Centaurs*, Florence, Casa Buonarroti. Photo Sopr. alle Gallerie, Florence
19. MICHELANGELO *Pietà*, Rome, St. Peter's. Photo Anderson
20. MICHELANGELO *Bacchus*, Florence, Bargello. Photo Sopr. alle Gallerie, Florence
21. FRA BARTOLOMMEO [with Albertinelli] *Last Judgment*, Florence, Museo di S. Marco. Photo Alinari

II
FORMATION OF THE CLASSICAL VOCABULARY

22. LEONARDO *St. Anne Cartoon*, London, Royal Academy. Photo Anderson
23. ANDREA DEL BRESCIANINO *St. Anne* (destroyed, formerly Berlin, Museums). Photo museum
24. MICHELANGELO *St. Anne*, Oxford, Ashmolean. Photo museum. Courtesy of the Visitors of the Ashmolean Museum
25. MICHELANGELO *Bruges Madonna*, Bruges, Notre Dame. Photo A. Breyne from Gevaert, Brussels
26. MICHELANGELO *David*, Florence, Academy. Photo Sopr. alle Gallerie, Florence
27. MICHELANGELO *Doni Holy Family*, Florence, Uffizi. Photo Anderson
28. LEONARDO *Battle of the Standard* (copy), Florence, Palazzo Vecchio. Photo Sopr. alle Gallerie, Florence
29. RUBENS *Battle of the Standard* (copy after Leonardo), Paris, Louvre. Photo museum
30. LEONARDO *Study for the Battle of Anghiari*, Venice, Academy. Photo Sopr. alle Gallerie, Florence
31. LEONARDO *Study for the Battle of Anghiari*, Venice, Academy. Photo Sopr. alle Gallerie, Florence
32. LEONARDO *Study for the Battle of Anghiari*, Budapest, Museum. Photo museum
33. LEONARDO *Study for the Battle of Anghiari*, Budapest, Museum. Photo museum
34. MICHELANGELO *St. Matthew*, Florence, Academy. Photo Alinari
35. MICHELANGELO *Study for the Julius Tomb* (incorporating the lower story of the project of 1505[?]; copy), Florence, Uffizi. Photo Sopr. alle Gallerie, Florence
36. MICHELANGELO *Battle of Cascina* (central portion; copy attributed to Aristotile da Sangallo), Holkham Hall. Photo Courtauld Institute, London
37. LEONARDO *Mona Lisa*, Paris, Louvre. Photo Alinari
38. LEONARDO *Mona Lisa* (detail). Photo Anderson
39. LEONARDO *St. Anne*, *Virgin*, *and Child*, Paris, Louvre. Photo Alinari
40. ALBERTINELLI *Madonna and Saints* (portable triptych), Milan, Poldi-Pezzoli. Photo Anderson
41. ALBERTINELLI *Crucifixion*, Florence, Certosa. Photo Sopr. alle Gallerie, Florence
42. ALBERTINELLI *Visitation*, Florence, Uffizi. Photo Brogi
43. FRA BARTOLOMMEO *Vision of St. Bernard*, Florence, Academy. Photo Sopr. alle Gallerie, Florence
44. FILIPPINO LIPPI *Vision of St. Bernard*, Florence, Badia. Photo Alinari
45. FRA BARTOLOMMEO *Noli Me Tangere*, Paris, Louvre. Photo Alinari

46. FRA BARTOLOMMEO [with Albertinelli] *Assumption of the Virgin* (destroyed; formerly Berlin, Museums). Photo museum
47. ALBERTINELLI *Annunciation with Sts. Sebastian and Lucy*, Munich, Pinakothek. Photo museum
48. RAPHAEL *Three Graces*, Chantilly, Musée Condé. Photo Alinari
49. RAPHAEL *Marriage of the Virgin*, Milan, Brera. Photo Anderson
50. RAPHAEL *Madonna del Granduca*, Florence, Pitti. Photo Sopr. alle Gallerie, Florence
51. RAPHAEL *Maddalena Doni*, Florence, Pitti. Photo Sopr. alle Gallerie, Florence
52. RAPHAEL *Angelo Doni*, Florence, Pitti. Photo Alinari
53. RAPHAEL *Madonna del Prato*, Vienna, Kunsthistorisches Museum. Photo museum.
54. RAPHAEL *Madonna del Cardellino*, Florence, Uffizi. Photo Sopr. alle Gallerie, Florence
55. RAPHAEL *La Belle Jardinière*, Paris, Louvre. Photo Alinari
56. RAPHAEL *Bridgewater House Madonna*, Edinburgh, National Gallery, Ellesmere Loan. Photo Annan, Glasgow. Courtesy of the Earl of Ellesmere
57. RAPHAEL *Madonna Study*, Paris, Louvre. Photo Alinari
58. RAPHAEL *Holy Family of the Casa Canigiani*, Munich, Pinakothek. Photo museum
59. RAPHAEL *Entombment*, Rome, Borghese. Photo Gabinetto Fotografico Nazionale, Rome
60. RAPHAEL *Madonna del Baldacchino*, Florence, Pitti. Photo Sopr. alle Gallerie, Florence
61. RAFFAELLINO CARLI *Madonna with Saints*, Florence, Sto. Spirito. Photo Alinari
62. LORENZO DI CREDI *Madonna with Saints*, Pistoia, S.M. delle Grazie. Photo Alinari
63. PIERO DI COSIMO *Madonna with Saints*, Florence, Innocenti. Photo Alinari
64. PIERO DI COSIMO *Immaculate Conception*, Florence, Uffizi. Photo Alinari
65. PIERO DI COSIMO *Madonna with St. John*, Vienna, Liechtenstein Collection. Photo Wolfrum
66. PIERO DI COSIMO *Liberation of Andromeda*, Florence, Uffizi 1536. Photo Anderson
67. GRANACCI *Holy Family*, Washington, National Gallery, Kress Collection. Photo Samuel H. Kress Collection, New York
68. GRANACCI *Madonna with Two Saints*, Berlin, Museums. Photo museum
69. GRANACCI *Holy Family with St. John*, Dublin, National Gallery. Photo museum. Courtesy of the National Gallery of Ireland
70. GRANACCI *Madonna della Cintola*, Florence, Academy. Photo Alinari

71. BUGIARDINI *Madonna and Child with St. John*, New York, Metropolitan Museum. Photo museum. Courtesy of the Metropolitan Museum of Art
72. BUGIARDINI *Holy Family*, Turin, Galleria Sabauda. Photo Alinari
73. BUGIARDINI *Portrait of a Lady*, Urbino, Galleria Nazionale. Photo Alinari
74. RIDOLFO GHIRLANDAIO *Madonna with Sts. Francis and Mary Magdalen*, Florence, Academy. Photo Alinari
75. RIDOLFO GHIRLANDAIO *Coronation of the Virgin*, Paris, Louvre. Photo Giraudon
76. RIDOLFO GHIRLANDAIO *Road to Calvary*, London, National Gallery. Photo museum. Reproduced by courtesy of the Trustees, The National Gallery, London
77. RIDOLFO GHIRLANDAIO *Marriage of St. Catherine*, Florence, Istituto delle Quiete. Photo Alinari
78. RIDOLFO GHIRLANDAIO *Altar Wing with Angels*, Florence, Academy. Photo Alinari
79. RIDOLFO GHIRLANDAIO *Altar Wing with Angels*, Florence, Academy. Photo Alinari
80. RIDOLFO GHIRLANDAIO *Portrait of a Man*, Chicago, Art Institute. Photo museum. Courtesy of the Art Institute, Chicago
81. RIDOLFO GHIRLANDAIO *Lady with a Rabbit*, New Haven, Yale University Art Gallery. Photo museum
82. RIDOLFO GHIRLANDAIO *Girl with a Unicorn*, Rome, Borghese. Photo Gabinetto Fotografico Nazionale, Rome
83. FRANCIABIGIO *Temple of Hercules*, Florence, Palazzo Davanzati. Photo Alinari
84. FRANCIABIGIO *Madonna and Child with St. John*, Florence, Uffizi 2178. Photo Brogi
85. FRANCIABIGIO *Madonna*, Perugia, Count Ranieri. Photo Fiorucci, Perugia, from Fototeca di Architettura e Topografia dell'Italia Antica, Rome
86. FRANCIABIGIO *Holy Family*, Florence, Academy (ex-Uffizi 888). Photo Alinari
87. FRANCIABIGIO *Madonna del Pozzo*, Florence, Academy. Photo Alinari
88. FRANCIABIGIO *Holy Family*, Vienna, Kunsthistorisches Museum 206. Photo museum
89. PINTURICCHIO *Choir Vault*, Rome, S.M. del Popolo. Photo Anderson
90. [Anon.] *Sibyls*, Rome, S. Pietro in Montorio. Photo Anderson
91. PERUZZI [and others] *Choir Decoration*, Rome, S. Onofrio. Photo Anderson
92. PERUZZI (?) *Coronation of the Virgin*, Rome, S. Onofrio. Photo Anderson
93. PERUZZI *Decoration of the Chapel*, Castello di Belcaro. Photo Alinari
94. PERUZZI *Madonna with Saints* (altar fresco), Rome, S. Onofrio. Photo Anderson

95. RIPANDA *Consul and Lictors*, Rome, Palazzo dei Conservatori. Photo Fototeca di Architettura e Topografia dell'Italia Antica, Rome
96. SODOMA *Central Octagon of the Stanza della Segnatura*, Rome, Vatican. Photo Alinari

III

THE MATURITY OF THE CLASSICAL STYLE IN ROME
(c. 1508–c. 1514)

97. MICHELANGELO *The Sistine Ceiling*, Rome, Vatican, Sistine Chapel. Photo Anderson
98. MICHELANGELO *The Sistine Ceiling* (central bays). Photo Archivio Fotografico delle Gallerie e Musei Vaticani
99. MICHELANGELO *Preliminary Plan for Sistine Ceiling*, London, British Museum. Photo museum. Courtesy of the Trustees of the British Museum
100. MICHELANGELO *Preliminary Plan for Sistine Ceiling*, Detroit, Institute of Arts. Photo museum. Courtesy of the Institute of Arts, Detroit
101. MICHELANGELO *Bronze-colored Nudes in the Spandrels*, Sistine Ceiling. Photo Archivio Fotografico delle Gallerie e Musei Vaticani
102. MICHELANGELO *Bronze-colored Nudes in the Spandrels*, Sistine Ceiling. Photo Archivio Fotografico delle Gallerie e Musei Vaticani
103. MICHELANGELO *The Flood*, Sistine Ceiling. Photo Archivio Fotografico delle Gallerie e Musei Vaticani
104. MICHELANGELO *Sacrifice of Noah*, Sistine Ceiling. Photo Anderson
105. MICHELANGELO *Drunkenness of Noah*, Sistine Ceiling. Photo Anderson
106. MICHELANGELO *Delphic Sibyl*, Sistine Ceiling. Photo Anderson
107. MICHELANGELO *Prophet Joel*, Sistine Ceiling. Photo Archivio Fotografico delle Gallerie e Musei Vaticani
108. MICHELANGELO *Prophet Zachary*, Sistine Ceiling. Photo Anderson
109. MICHELANGELO *Erithrean Sibyl*, Sistine Ceiling. Photo Anderson
110. MICHELANGELO *Prophet Isaiah*, Sistine Ceiling. Photo Anderson
111. MICHELANGELO *Ignudi around the Drunkennesss of Noah*, Sistine Ceiling. Photo Alinari
112. MICHELANGELO *Ignudo above Prophet Joel*, Sistine Ceiling. Photo Anderson
113. MICHELANGELO *Ignudo above Prophet Joel*, Sistine Ceiling. Photo Anderson
114. MICHELANGELO *Ignudi around Sacrifice of Noah*, Sistine Ceiling. Photo Alinari
115. MICHELANGELO *Ignudo above Prophet Isaiah*, Sistine Ceiling. Photo Anderson
116. MICHELANGELO *Ignudo above Prophet Isaiah*, Sistine Ceiling. Photo Anderson

117. MICHELANGELO *Creation of Eve*, Sistine Ceiling. Photo Archivio Fotografico delle Gallerie e Musei Vaticani
118. MICHELANGELO *Ignudi around Creation of Eve*, Sistine Ceiling. Photo Alinari
119. MICHELANGELO *Ignudo above Cumaean Sibyl*, Sistine Ceiling. Photo Anderson
120. MICHELANGELO *Ignudo above Cumaean Sibyl*, Sistine Ceiling. Photo Anderson
121. MICHELANGELO *Temptation and Expulsion*, Sistine Ceiling. Photo Archivio Fotografico delle Gallerie e Musei Vaticani
122. MICHELANGELO *Creation of Adam*, Sistine Ceiling. Photo Archivio Fotografico delle Gallerie e Musei Vaticani
123. MICHELANGELO *Creation of Adam* (detail), Sistine Ceiling. Photo Anderson
124. MICHELANGELO *Creation of the Sun and Moon*, Sistine Ceiling. Photo Archivio Fotografico delle Gallerie e Musei Vaticani
125. MICHELANGELO *Separation of Earth and Waters*, Sistine Ceiling. Photo Anderson
126. MICHELANGELO *Separation of Light and Darkness*, Sistine Ceiling. Photo Anderson
127. MICHELANGELO *Ignudi around Separation of Earth and Waters*, Sistine Ceiling. Photo Alinari
128. MICHELANGELO *Ignudo above Persian Sibyl*, Sistine Ceiling. Photo Anderson
129. MICHELANGELO *Ignudo above Persian Sibyl*, Sistine Ceiling. Photo Anderson
130. MICHELANGELO *Ignudi around Separation of Light and Darkness*, Sistine Ceiling. Photo Anderson
131. MICHELANGELO *Ignudo above Prophet Jeremiah*, Sistine Ceiling. Photo Anderson
132. MICHELANGELO *Ignudo above Prophet Jeremiah*, Sistine Ceiling. Photo Anderson
133. MICHELANGELO *Cumaean Sibyl*, Sistine Ceiling. Photo Anderson
134. MICHELANGELO *Persian Sibyl*, Sistine Ceiling. Photo Archivio Fotografico delle Gallerie e Musei Vaticani
135. MICHELANGELO *Prophet Ezekiel*, Sistine Ceiling. Photo Anderson
136. MICHELANGELO *Prophet Daniel*, Sistine Ceiling. Photo Archivio Fotografico delle Gallerie e Musei Vaticani
137. MICHELANGELO *Libyan Sibyl*, Sistine Ceiling. Photo Archivio Fotografico delle Gallerie e Musei Vaticani
138. MICHELANGELO *Prophet Jonah*, Sistine Ceiling. Photo Archivio Fotografico delle Gallerie e Musei Vaticani
139. MICHELANGELO *Prophet Jeremiah*, Sistine Ceiling. Photo Anderson

140. MICHELANGELO *David and Goliath*, Sistine Ceiling. Photo Anderson
141. MICHELANGELO *Judith and Holofernes*, Sistine Ceiling. Photo Anderson
142. MICHELANGELO *The Hanging of Haman*, Sistine Ceiling. Photo Anderson
143. MICHELANGELO *The Brazen Serpent*, Sistine Ceiling. Photo Anderson
144. MICHELANGELO *Ancestors of Christ*, Sistine Ceiling, severies. Photo Anderson
145. MICHELANGELO *Ancestors of Christ*, Sistine Ceiling, severies. Photo Anderson
146. MICHELANGELO *Ancestors of Christ*, Sistine Ceiling, lunettes. Photo Alinari
147. MICHELANGELO *Ancestors of Christ*, Sistine Ceiling, lunettes. Photo Alinari
148. MICHELANGELO *Ancestors of Christ*, Sistine Ceiling. Photo Anderson
149. RAPHAEL *Stanza della Segnatura*, Rome, Vatican. Photo Gabinetto Fotografico Nazionale, Rome
150. RAPHAEL *Stanza della Segnatura*. Photo Alinari
151. RAPHAEL [and Sodoma] *Ceiling of the Stanza della Segnatura*. Photo Anderson
152. RAPHAEL *Poetry*, Stanza della Segnatura, ceiling. Photo Archivio Fotografico delle Gallerie e Musei Vaticani
153. RAPHAEL *Justice*, Stanza della Segnatura, ceiling. Photo Archivio Fotografico delle Gallerie e Musei Vaticani
154. RAPHAEL *The Flaying of Marsyas*, Stanza della Segnatura, ceiling. Photo Anderson
155. RAPHAEL *The Judgment of Solomon*, Stanza della Segnatura, ceiling. Photo Anderson
156. RAPHAEL *Disputà*, Stanza della Segnatura. Photo Anderson
157. RAPHAEL *Disputà* (detail). Photo Anderson
158. RAPHAEL *Disputà* (detail). Photo Anderson
159. RAPHAEL *Parnassus*, Stanza della Segnatura. Photo Archivio Fotografico delle Gallerie e Musei Vaticani
160. RAPHAEL *Parnassus* (detail), *Apollo and Muses*. Photo Anderson
161. RAPHAEL *Parnassus* (detail), *Sappho and Other Poets*. Photo Anderson
162. RAPHAEL *Parnassus* (detail), *Modern Poets*. Photo Archivio Fotografico delle Gallerie e Musei Vaticani
163. RAPHAEL *School of Athens*, Stanza della Segnatura. Photo Archivio Fotografico delle Gallerie e Musei Vaticani
164. RAPHAEL *School of Athens* (detail), *Plato and Aristotle*. Photo Achivio Fotografico delle Gallerie e Musei Vaticani
165. RAPHAEL *School of Athens* (detail), *Pythagorean Group*. Photo Archivio Fotografico delle Gallerie e Musei Vaticani
166. RAPHAEL *School of Athens* (detail), *Euclidian Group*. Photo Archivio Fotografico delle Gallerie e Musei Vaticani

167. RAPHAEL *School of Athens* (detail), *Heraclitus*. Photo Archivio Fotografico delle Gallerie e Musei Vaticani
168. RAPHAEL *Cartoon for School of Athens* (detail), Milan, Ambrosiana. Photo Anderson
169. RAPHAEL *Cartoon for School of Athens*, Milan, Ambrosiana. Photo Alinari
170. RAPHAEL *The Law*, Stanza della Segnatura. Photo Archivio Fotografico delle Gallerie e Musei Vaticani
171. RAPHAEL *Three Virtues of Justice*, Stanza della Segnatura. Photo Archivio Fotografico delle Gallerie e Musei Vaticani
172. RAPHAEL *Handing over of the Decretals*, Stanza della Segnatura. Photo Archivio Fotografico delle Gallerie e Musei Vaticani
173. RAPHAEL *The Civil Law*, Stanza della Segnatura. Photo Anderson
174. RAPHAEL [with Penni] *Grisaille and Grotesque Decoration* (beneath *Parnassus*), Stanza della Segnatura. Photo Archivio Fotografico delle Gallerie e Musei Vaticani
175. RAPHAEL *Madonna di Casa Alba*, Washington, National Gallery (Mellon Collection). Photo museum
176. RAPHAEL *Study for Alba Madonna*, Lille, Musée Wicar. Photo Gérondal, Lille
177. RAPHAEL *Madonna di Foligno*, Rome, Vatican Museum. Photo Archivio Fotografico delle Gallerie e Musei Vaticani
178. RAPHAEL *Isaiah*, Rome, S. Agostino. Photo Gabinetto Fotografico Nazionale, Rome
179. RIPANDA (?) *Triumph of Titus*, Paris, Louvre 180. Photo museum
180. [Anon.] *Façade Decoration of the Casa Sander*, Rome. Photo Fototeca di Architettura e Topografia dell'Italia Antica, Rome
181. [Anon.] *Façade Decoration*, Via Maschera d'Oro no. 9, Rome. Photo Fototeca di Architettura e Topografia dell'Italia Antica, Rome
182. PERUZZI *Three Graces*, San Francisco, Zellerbach Collection. Photo Moulin Studios, San Francisco. Courtesy of Mr. J. D. Zellerbach
183. PERUZZI *Sala di Galatea*, Rome, Villa Farnesina. Photo Fototeca di Architettura e Topografia dell'Italia Antica, Rome
184. PERUZZI *Aquarius between the Swan and Dolphin.* SEBASTIANO *Tireus and Philomel; The Daughters of Cecrops*, Sala di Galatea. Photo Alinari
185. PERUZZI *Perseus-Pegasus*, Sala di Galatea. Photo Alinari
186. PERUZZI *Ursa Major*, Sala di Galatea. Photo Alinari
187. PERUZZI *Luna in Virgo, Bacchus and Ariadne, Mars in Libra near Scorpio* SEBASTIANO *Fall of Phaeton*, Sala di Galatea. Photo Alinari
188. PERUZZI *Venus in Capricorn with Sagittarius and Lyra*, Sala di Galatea. Photo Fototeca di Architettura e Topografia dell'Italia Antica, Rome
189. PERUZZI *Argo*, Sala di Galatea. Photo Fototeca di Architettura e Topografia dell'Italia Antica, Rome
190. PERUZZI *Sol in Sagittarius*, Sala di Galatea. Photo Anderson

191. PERUZZI *Death of Meleager*, Rome, Villa Farnesina, Sala del Fregio. Photo Gabinetto Fotografico Nazionale, Rome
192. PERUZZI *Hunting of Calydonian Boar*, Sala del Fregio. Photo Gabinetto Fotografico Nazionale, Rome
193. PERUZZI *Nymph and Satyrs; Slaying of Marsyas* (portions), Sala del Fregio. Photo Gabinetto Fotografico Nazionale, Rome
194. SODOMA *Marriage of Alexander and Roxane*, Rome, Villa Farnesina. Photo Gabinetto Fotografico Nazionale, Rome
195. SODOMA *Alexander and the Family of Darius; The Forge of Vulcan*, Rome, Villa Farnesina. Photo Alinari
196. SEBASTIANO *Juno*, Sala di Galatea. Photo Anderson
197. SEBASTIANO *Fall of Icarus*, Sala di Galatea. Photo Anderson
198. SEBASTIANO *Polyphemus*, Sala di Galatea. Photo Fototeca di Architettura e Topografia dell'Italia Antica, Rome
199. SEBASTIANO *Adoration of the Shepherds*, Cambridge, Fitzwilliam Museum. Photo Stearn and Sons. Reproduced by permission of the Syndics of the Fitzwilliam Museum, Cambridge
200. SEBASTIANO *Death of Adonis*, Florence, Uffizi. Photo Alinari
201. SEBASTIANO *Madonna and Child*, London, Pouncey Collection. Photo The National Gallery, London. Reproduced by courtesy of Mr. Philip Pouncey, London
202. SEBASTIANO *Portrait of a Girl* (called "'La Fornarina"), Florence, Uffizi. Photo Anderson
203. SEBASTIANO *Portrait of a Girl* (called "Dorothea"), Berlin, Museums. Photo museum
204. PERUZZI *Giant Head*, Sala di Galatea, Rome, Villa Farnesina. Photo Gabinetto Fotografico Nazionale, Rome
205. PERUZZI *Ceiling of the Stanza d'Eliodoro*, Rome, Vatican. Photo Anderson
206. PERUZZI (?) *Project for Stanza d'Eliodoro*, Paris, Louvre. Photo museum
207. PERUZZI *Study for Ceiling*, Stanza d'Eliodoro; Oxford, Ashmolean. Photo museum. Courtesy of the Visitors of the Ashmolean Museum
208. PERUZZI *Cartoon for Moses and the Burning Bush*, Stanza d'Eliodoro; Naples, Galleria Nazionale. Photo Alinari
209. PERUZZI *Moses and the Burning Bush*, Stanza d'Eliodoro, ceiling. Photo Anderson
210. PERUZZI *Jacob's Dream*, Stanza d'Eliodoro, ceiling. Photo Anderson
211. PERUZZI *Adoration of the Child*, Rome, S. Rocco. Photo Anderson
212. PERUZZI *Holy Family in a Landscape*, London, Pouncey Collection. Photo courtesy of Mr. Philip Pouncey, London
213. RAPHAEL *Stanza d'Eliodoro*, Rome Vatican. Photo Alinari
214. RAPHAEL *Mass of Bolsena*, Stanza d'Eliodoro. Photo Anderson
215. RAPHAEL *Mass of Bolsena* (detail). Photo Gabinetto Fotografico Nazionale, Rome

216. RAPHAEL *Mass of Bolsena* (detail), *Swiss Guards*. Photo Archivio Fotografico delle Gallerie e Musei Vaticani
217. RAPHAEL *Expulsion of Heliodorus*, Stanza d'Eliodoro. Photo Anderson
218. RAPHAEL *Expulsion of Heliodorus* (detail). Photo Alinari
219. RAPHAEL *Expulsion of Heliodorus* (detail). Photo Archivio Fotografico delle Gallerie e Musei Vaticani
220. RAPHAEL *Repulse of Attila* (detail). Photo Archivio Fotografico delle Gallerie e Musei Vaticani
221. RAPHAEL *Repulse of Attila*, Stanza d'Eliodoro. Photo Anderson
222. RAPHAEL *Liberation of Peter*, Stanza d'Eliodoro. Photo Anderson
223. RAPHAEL *Liberation of Peter* (detail). Photo Archivio Fotografico delle Gallerie e Musei Vaticani
224. RAPHAEL *Liberation of Peter* (detail). Photo Archivio Fotografico delle Gallerie e Musei Vaticani
225. RAPHAEL ASSISTANT *Grotesque Decoration*, Stanza d'Eliodoro. Photo Alinari
226. GIULIO *Basamento* (detail; repainted), Stanza d'Eliodoro. Photo Archivio Fotografico delle Gallerie e Musei Vaticani
227. RAPHAEL *Study for a Resurrection*, Bayonne, Musée Bonnat. Photo Archives Photographiques, Paris
228. RAPHAEL *Study for a Resurrection*, Windsor, Royal Library. Photo Royal Library. Copyright reserved
229. RAPHAEL *Study for a Resurrection*, Oxford, Ashmolean. Photo museum. Courtesy of the Visitors of the Ashmolean Museum
230. RAPHAEL *Galatea*, Rome, Villa Farnesina. Photo Gabinetto Fotografico Nazionale, Rome
231. RAPHAEL ASSISTANT *Prophets*, Rome, S.M. della Pace, Cappella Chigi. Photo Gabinetto Fotografico Nazionale, Rome
232. RAPHAEL *Sibyls*, S.M. della Pace, Cappella Chigi. Photo Alinari
233. RAPHAEL *Sibyls* (detail). Photo Anderson
234. RAPHAEL *Sibyls* (detail). Photo Anderson
235. RAPHAEL *Sistine Madonna*, Dresden, Gallery. Photo Alinari
236. RAPHAEL *Sistine Madonna* (detail), *St. Sixtus*. Photo museum
237. RAPHAEL *Sistine Madonna* (detail), *St. Barbara*. Photo museum
238. RAPHAEL [with Penni and Giulio] *Madonna of the Fish*, Madrid, Prado. Photo Anderson
239. RAPHAEL *Study for Madonna dell'Impannata*, Windsor, Royal Library. Photo Royal Library. Copyright reserved
240. RAPHAEL [with Giulio] *Madonna dell'Impannata*, Florence, Pitti. Photo Alinari
241. RAPHAEL *St. Cecilia Altar*, Bologna, Pinacoteca. Photo Alinari
242. RAPHAEL *St. Cecilia Altar* (detail). Photo Edizioni Cartovendita, Bologna

243. RAPHAEL *St. Cecilia Altar* (detail). Photo Edizioni Cartovendita, Bologna
244. RAPHAEL *Tommaso Inghirami* ("Il Fedra"), Boston, Isabella Stewart Gardner Museum. Photo museum
245. RAPHAEL ASSISTANT *Tommaso Inghirami*, Florence, Pitti. Photo Anderson
246. RAPHAEL [or assistant] *Giuliano de' Medici*, New York, Metropolitan Museum. Photo museum. Courtesy of the Metropolitan Museum of Art
247. RAPHAEL *La Donna Velata*, Florence, Pitti. Photo Sopr. alle Gallerie, Florence
248. RAPHAEL *Madonna della Sedia*, Florence, Pitti. Photo Sopr. alle Gallerie, Florence
249. RAPHAEL *Madonna della Sedia* (detail). Photo Alinari
250. LEONARDO [and assistant?] *St. John Baptist*, Paris, Louvre. Photo Alinari
251. LEONARDO *Cataclysm*, Windsor, Royal Library. Photo Royal Library. Copyright reserved

IV

THE MATURATION OF CLASSICAL STYLE IN FLORENCE
(*c.* 1508–*c.* 1514)

252. FRA BARTOLOMMEO *Holy Family*, London, National Gallery 3914. Photo museum. Reproduced by courtesy of the Trustees, The National Gallery, London
253. FRA BARTOLOMMEO *God the Father with Sts. Mary Magdalen and Catherine of Siena*, Lucca, Pinacoteca. Photo Alinari
254. FRA BARTOLOMMEO [with Albertinelli] *Madonna with Six Saints*, Florence, S. Marco. Photo Alinari
255. FRA BARTOLOMMEO *Madonna with Sts. Stephen and John Baptist*, Lucca, Cathedral. Photo Alinari
256. ALBERTINELLI *Madonna with Four Saints*, Florence, Academy. Photo Alinari
257. ALBERTINELLI *Trinity*, Florence, Academy. Photo Sopr. alle Gallerie, Florence
258. ALBERTINELLI *Study for a Trinity*, Florence, Uffizi. Photo Sopr. alle Gallerie, Florence
259. ALBERTINELLI *Annunciation*, Florence, Academy. Photo Alinari
260. FRA BARTOLOMMEO *Marriage of St. Catherine*, Paris, Louvre. Photo Alinari
261. FRA BARTOLOMMEO [with Albertinelli] *Virgin in Glory with Saints*, Besançon, Cathedral. Photo Bulloz, Paris
262. ALBERTINELLI *Coronation of the Virgin* (fragment; former crown-piece of 261), Stuttgart, Gallery. Photo museum

263. FRA BARTOLOMMEO *St. Anne Altar*, Florence, Museo di S. Marco. Photo Sopr. alle Gallerie, Florence
264. FRA BARTOLOMMEO [with assistants] *The Marriage of St. Catherine* (the *Pitti Pala*), Florence, Academy. Photo Alinari
265. FRA BARTOLOMMEO [with Raphael] *St. Peter*, Rome, Vatican Museum. Photo Alinari
266. FRA BARTOLOMMEO *St. Paul*, Rome, Vatican Museum. Photo Alinari
267. FRA BARTOLOMMEO [with Raphael] *St. Peter* (detail). Photo Anderson
268. FRA BARTOLOMMEO *Study for St. Paul*, Florence, Uffizi. Photo Sopr. alle Gallerie, Florence
269. BUGIARDINI *La Monaca*, Florence, Pitti. Photo Alinari
270. BUGIARDINI *Portrait of a Young Woman*, Paris, Musée Jacquemart-André. Photo Bulloz, Paris
271. BUGIARDINI *Madonna Standing in a Landscape* (sold London, 1946). Photo A. C. Cooper
272. BUGIARDINI *Madonna and Child with St. John* (formerly?) New York, C. H. Holmes Collection. Photo Gray
273. BUGIARDINI *Madonna and Child with St. John* (formerly) London, Agnew. Photo courtesy of the Frick Art Reference Library, New York
274. BUGIARDINI *Madonna del Latte*, Florence, Uffizi. Photo Alinari
275. BUGIARDINI *Ariadne* (?), Venice, Ca' d'Oro. Photo Alinari
276. BUGIARDINI *Leda*, Milan, Treccani Collection. Photo Alinari
277. RIDOLFO GHIRLANDAIO *Portrait of a Lady*, Florence, Pitti. Photo Sopr. alle Gallerie, Florence
278. RIDOLFO GHIRLANDAIO *Adoration of the Child* (destroyed; formerly Berlin, Museums). Photo museum
279. RIDOLFO GHIRLANDAIO *Adoration of the Shepherds*, Budapest, Museum. Photo museum
280. RIDOLFO GHIRLANDAIO *Nativity with Six Saints*, New York, Metropolitan Museum. Photo Anderson
281. RIDOLFO GHIRLANDAIO [shop assistant] *Adoration of the Shepherds* (formerly) London, Henry Harris Collection. Photo A. C. Cooper
282. RIDOLFO GHIRLANDAIO *Madonna della Cintola*, Prato, Cathedral. Photo Alinari
283. RIDOLFO GHIRLANDAIO *Portrait of a Goldsmith*, Florence, Pitti. Photo Sopr. alle Gallerie, Florence
284. RIDOLFO GHIRLANDAIO [with Andrea di Cosimo] *Decoration of Cappella dei Priori*, Florence, Palazzo Vecchio. Photo Alinari
285. RIDOLFO GHIRLANDAIO [with Andrea di Cosimo] *Decoration of Cappella dei Priori*. Photo Alinari
286. GRANACCI *Madonna with Two Saints*, Villamagna (near Florence), S. Donnino. Photo courtesy of Frick Art Reference Library, New York

287. GRANACCI *Madonna della Cintola*, Sarasota, Ringling Museum. Photo museum
288. GRANACCI *Trinity*, Berlin, Museums. Photo museum
289. GRANACCI *Pietà*, Quintole, S. Pietro. Photo Brogi
290. GRANACCI *Madonna in Glory with Four Saints*, Florence, Academy. Photo Alinari
291. GRANACCI *Holy Family with St. John*, Florence, Pitti. Photo Brogi
292. PIERO DI COSIMO [with assistants] *Doctrine of the Immaculate Conception*, Fiesole, S. Francesco. Photo Alinari
293. PIERO DI COSIMO *Legend of Prometheus*, Munich, Pinakothek. Photo museum
294. PIERO DI COSIMO *Legend of Prometheus*, Strasbourg, Museum. Photo museum
295. PIERO DI COSIMO *Adoration of the Child*, Rome, Borghese. Photo Archivio Fotografico delle Gallerie e Musei Vaticani
296. ANDREA DEL SARTO *Pietà*, Rome, Borghese. Photo Anderson
297. ANDREA DEL SARTO *Madonna*, Rome, Galleria Nazionale. Photo Alinari
298. ANDREA DEL SARTO *Baptism of Christ*, Florence, Scalzo. Photo Alinari
299. ANDREA DEL SARTO *Healing of the Obsessed Girl*, Florence, SS. Annunziata. Photo Alinari
300. ANDREA DEL SARTO *Burial of St. Philip*, Florence, SS. Annunziata. Photo Alinari
301. ANDREA DEL SARTO *Healing by St. Philip's Relics*, Florence, SS. Annunziata. Photo Alinari
302. ANDREA DEL SARTO *Punishment of the Gamblers*, Florence, SS. Annunziata. Photo Alinari
303. ANDREA DEL SARTO *Clothing of the Leper*, Florence, SS. Annunziata. Photo Alinari
304. ANDREA DEL SARTO *Noli Me Tangere*, Florence, Uffizi. Photo Alinari
305. FRANCIABIGIO *Madonna*, Rome, Galleria Nazionale. Photo Gabinetto Fotografico Nazionale, Rome
306. FRANCIABIGIO *Adoration of the Shepherds*, Florence, Museo di S. Marco. Photo Sopr. alle Gallerie, Florence
307. FRANCIABIGIO *Last Supper*, Florence, S.M. dei Candeli. Photo Brogi
308. ANDREA DEL SARTO *Adoration of the Magi*, Florence, SS. Annunziata. Photo Alinari
309. ANDREA DEL SARTO *Annunciation*, Florence, Pitti. Photo Alinari
310. ANDREA DEL SARTO [with Puligo?] *Marriage of St. Catherine*, Dresden, Gallery. Photo Alinari
311. ANDREA DEL SARTO *Birth of the Virgin*, Florence, SS. Annunziata. Photo Sopr. alle Gallerie, Florence

312. ANDREA DEL SARTO *Birth of the Virgin* (detail). Photo Alinari
313. ANDREA DEL SARTO *Birth of the Virgin* (detail). Photo Sopr. alle Gallerie, Florence
314. FRANCIABIGIO *Marriage of the Virgin*, Florence, SS. Annunziata. Photo Alinari
315. FRANCIABIGIO *Last Supper*, Florence, Convento della Calza. Photo Sopr. alle Gallerie, Florence
316. FRANCIABIGIO *Last Supper* (detail). Photo Sopr. alle Gallerie, Florence
317. FRANCIABIGIO *Portrait of a Man*, Florence, Uffizi 8381. Photo Alinari
318. FRANCIABIGIO *Portrait of a Man*, London, National Gallery. Photo museum. Reproduced by courtesy of the Trustees, The National Gallery, London
319. PULIGO *Madonna and Child with St. John*, Rome, Palazzo Venezia. Photo Gabinetto Fotografico Nazionale, Rome
320. ANDREA DEL SARTO [with Puligo?] *Madonna with the Infant St. John*, Rome, Borghese 336. Photo Alinari
321. ANDREA DEL SARTO [with Puligo?] *Tobias Altar*, Vienna, Kunsthistorisches Museum. Photo museum
322. PULIGO *Madonna with St. John Approaching in a Landscape*, Rome, Borghese 338. Photo Gabinetto Fotografico Nazionale, Rome
323. PONTORMO *Ospedale di S. Matteo*, Florence, Academy. Photo Brogi
324. PONTORMO *Madonna with Four Saints* (altar fresco from S. Ruffillo), Florence, SS. Annunziata. Photo Sopr. alle Gallerie, Florence
325. ROSSO *Madonna and Child*, New York, Finch College, Kress Collection 485. Photo Samuel H. Kress Collection, New York
326. ROSSO *Madonna in a Landscape*, Arezzo, Museum (from Uffizi Deposit 8309). Photo author
327. ROSSO *Holy Family*, Rome, Borghese. Photo Gabinetto Fotografico Nazionale, Rome
328. BERRUGUETE *Madonna and Elizabeth with the Two Holy Children*, Rome, Borghese 335. Photo Gabinetto Fotografico Nazionale, Rome
329. BERRUGUETE *Madonna*, Milan, Saibene Collection. Photo courtesy of Dr. Alberto Saibene, Milan
330. [Anon.] *Madonna* (formerly) Milan, Crespi Collection. Photo Anderson
331. FILIPPINO LIPPI, BERRUGUETE [and others] *Coronation of the Virgin*, Paris, Louvre. Photo Alinari
332. MANCHESTER MASTER *Madonna and Child with St. John*, Vienna, Academy. Photo museum
333. MANCHESTER MASTER *Virgin Reading with the Christ Child and St. John*, New York, Kress Collection, Photo Samuel H. Kress Collection, New York
334. MANCHESTER MASTER *Pietà*, Rome, Galleria Nazionale. Photo Gabinetto Fotografico Nazionale, Rome

335. MANCHESTER MASTER   *Madonna*, Baden bei Zürich, Private Collection. Photo courtesy of Prof. Federico Zeri, Rome
336. MANCHESTER MASTER   *Madonna with St. John and Four Angels*, London, National Gallery. Photo museum. Reproduced by courtesy of the Trustees, The National Gallery, London
337. MANCHESTER MASTER   *Madonna*, Florence, Art Market. Photo courtesy of Prof. Federico Zeri, Rome
338. MICHELANGELO [with the Manchester Master]   *Entombment*, London National Gallery. Photo museum. Reproduced by courtesy of the Trustees, The National Gallery, London

V

CLIMAX, CRISIS, AND DISSOLUTION OF THE CLASSICAL STYLE IN ROME (*c.* 1514–*c.* 1520)

339. RAPHAEL   *Acts of the Apostles* (tapestries), Rome, Vatican Museum (in order of arrangement in the Sistine Chapel). Photos Alinari
340. RAPHAEL   *The Stoning of Stephen* (tapestry). Photo Alinari
341. RAPHAEL   *The Conversion of Paul* (tapestry). Photo Alinari
342. RAPHAEL   *Miraculous Draught of Fishes* (tapestry). Photo Alinari
343. RAPHAEL   *Pasce Oves* (tapestry). Photo Alinari
344. RAPHAEL   *Blinding of Elymas* (tapestry, fragment). Photo Alinari
345. RAPHAEL   *Paul at Lystra* (tapestry). Photo Alinari
346. RAPHAEL   *Healing at the Golden Gate* (tapestry). Photo Alinari
347. RAPHAEL   *Death of Ananias* (tapestry). Photo Anderson
348. RAPHAEL   *Paul Preaching at Athens* (tapestry). Photo Alinari
349. RAPHAEL   *Study for the Pasce Oves*, Windsor, Royal Library. Photo Royal Library. Copyright reserved
350. RAPHAEL   *Study for the Christ of Pasce Oves*, Paris, Louvre. Photo museum
351. RAPHAEL   *Study for the Paul at Lystra*, Chatsworth. Photo Devonshire Collection, Chatsworth. Reproduced by permission the Trustees of the Chatsworth Settlement
352. RAPHAEL   *Study for the Blinding of Elymas*, Windsor, Royal Library. Photo Royal Library. Copyright reserved
353. PENNI   *Study for the Pasce Oves*, Paris, Louvre. Photo museum
354. PENNI   *Study for Paul Preaching at Athens*, Florence, Uffizi. Photo Sopr. alle Gallerie, Florence
355. RAPHAEL   *Miraculous Draught of Fishes* (tapestry cartoon), London, Victoria and Albert Museum. Photo museum
356. RAPHAEL   *Miraculous Draught of Fishes* (tapestry cartoon, detail). Photo museum
357. RAPHAEL   *Miraculous Draught of Fishes* (tapestry cartoon, detail). Photo museum

358. RAPHAEL *Pasce Oves* (tapestry cartoon), London, Victoria and Albert Museum. Photo museum
359. RAPHAEL *Healing at the Golden Gate* (tapestry cartoon), London, Victoria and Albert Museum. Photo museum
360. RAPHAEL *Healing at the Golden Gate* (tapestry cartoon, detail). Photo museum
361. RAPHAEL *Healing at the Golden Gate* (tapestry cartoon, detail). Photo museum
362. RAPHAEL *Death of Ananias* (tapestry cartoon), London, Victoria and Albert Museum. Photo museum
363. RAPHAEL *Death of Ananias* (tapestry cartoon, detail). Photo museum
364. RAPHAEL *Death of Ananias* (tapestry cartoon, detail). Photo museum
365. RAPHAEL *Death of Ananias* (tapestry cartoon, detail). Photo museum
366. RAPHAEL [with Penni] *Blinding of Elymas* (tapestry cartoon), London, Victoria and Albert Museum. Photo museum
367. RAPHAEL *Paul at Lystra* (tapestry cartoon), London, Victoria and Albert Museum. Photo museum
368. RAPHAEL *Paul at Lystra* (tapestry cartoon, detail). Photo museum
369. RAPHAEL [with Penni] *Paul Preaching at Athens* (tapestry cartoon), London, Victoria and Albert Museum. Photo museum
370. PENNI *Paul Preaching at Athens* (tapestry cartoon, detail). Photo museum
371. RAPHAEL *Paul at Lystra* (tapestry cartoon, detail). Photo museum
372. PENNI *Study for the Miraculous Draught of Fishes*, Vienna, Albertina. Photo museum
373. GIULIO [retouched by Raphael?] *Study for the Battle of Ostia*, Vienna, Albertina. Photo museum
374. RAPHAEL [with Giulio] *Battle of Ostia*, Rome, Vatican, Stanza dell'Incendio. Photo Alinari
375. GIULIO *Battle of Ostia* (detail). Photo courtesy of Frick Art Reference Library, New York
376. RAPHAEL [with Giulio] *Fire in the Borgo*, Rome, Vatican, Stanza dell' Incendio. Photo Anderson
377. GIULIO *Fire in the Borgo* (detail). Photo Gabinetto Fotografico Nazionale, Rome
378. GIULIO *Fire in the Borgo* (detail). Photo Archivio Fotografico delle Gallerie e Musei Vaticani
379. GIULIO *Fire in the Borgo* (detail). Photo Archivio Fotografico delle Gallerie e Musei Vaticani
380. RAPHAEL [with Penni] *Coronation of Charlemagne*, Rome, Vatican, Stanza dell'Incendio. Photo Anderson
381. PENNI *Coronation of Charlemagne* (detail). Photo Gabinetto Fotografico Nazionale, Rome

382. PENNI *Study for the Coronation of Charlemagne*, Düsseldorf, Museum. Photo Landesbildstelle Rheinland
383. RAPHAEL [with Penni] *Oath of Leo*, Rome, Vatican, Stanza dell'Incendio. Photo Anderson
384. PENNI *Study for the Oath of Leo*, Florence, Horne Foundation. Photo Sopr. alle Gallerie, Florence
385. GIULIO *Basamento* (detail), Rome, Vatican, Stanza dell'Incendio. Photo Alinari
386. RAPHAEL [with Alvise de Pace] *Cupola of the Cappella Chigi*, Rome, S.M. del Popolo. Photo Alinari
387. RAPHAEL *Study for the Cappella Chigi*, Oxford, Ashmolean. Photo museum. Courtesy of the Visitors of the Ashmolean Museum
388. RAPHAEL *Study for the Cappella Chigi*, Oxford, Ashmolean. Photo museum. Courtesy of the Visitors of the Ashmolean Museum
389. RAPHAEL [with Giovanni da Udine] *Loggetta of the Cardinal Bibbiena*, Rome, Vatican. Photo Archivio Fotografico delle Gallerie e Musei Vaticani
390. RAPHAEL [with Giovanni da Udine] *Loggetta of the Cardinal Bibbiena*. Photo Archivio Fotografico delle Gallerie e Musei Vaticani
391. GIOVANNI DA UDINE [and assistants] *Loggetta of the Cardinal Bibbiena* (detail). Photo Archivio Fotografico delle Gallerie e Musei Vaticani
392. RAPHAEL [with Giovanni da Udine] *Stufetta of the Cardinal Bibbiena*, Rome, Vatican. Photo Archivio Fotografico delle Gallerie e Musei Vaticani
393. RAPHAEL [with Giovanni da Udine] *Stufetta of the Cardinal Bibbiena*. Photo Archivio Fotografico delle Gallerie e Musei Vaticani
394. TADDEO ZUCCARO [and others] *Decoration of the Sala dei Palafrenieri* (free reconstruction of Raphael), Rome, Vatican. Photo Alinari
395. PENNI *Study for the Sala dei Palafrenieri*, Paris, Louvre. Photo museum
396. PENNI *Study for the Sala dei Palafrenieri*, Paris, Louvre. Photo museum
397. RAPHAEL [and assistants] *Sala di Psiche*, Rome, Villa Farnesina. Photo Fototeca di Architettura e Topografia dell'Italia Antica, Rome
398. GIULIO *Three Graces*, Sala di Psiche. Photo Gabinetto Fotografico Nazionale, Rome
399. GIULIO *Venus, Ceres, and Juno*, Sala di Psiche. Photo Alinari
400. GIULIO *Jupiter and Cupid*, Sala di Psiche. Photo Alinari
401. GIULIO *Venus and Psyche*, Sala di Psiche. Photo Alinari
402. RAPHAEL *Study for Venus and Psyche*, Paris, Louvre. Photo museum
403. PENNI *Venus before Jupiter*, Rome, Villa Farnesina, Sala di Psiche. Photo Alinari
404. PENNI *Mercury Descending*, Sala di Psiche. Photo Alinari
405. RAFFAELLINO DEL COLLE (?) *Venus and Cupid*, Sala di Psiche. Photo Alinari
406. GIOVANNI DA UDINE *Amoretto with Mythic Beasts*, Sala di Psiche. Photo Alinari

407. RAPHAEL *Study for the Wedding Feast of Cupid and Psyche*, Windsor, Royal Library. Photo Royal Library, Windsor
408. PENNI *Council of the Gods*, Rome, Villa Farnesina, Sala di Psiche. Photo Alinari
409. PENNI *Wedding Feast of Cupid and Pysche*, Sala di Psiche. Photo Alinari
410. RAPHAEL [and assistants] *Logge*, Rome, Vatican. Photo Archivio Fotografico delle Gallerie e Musei Vaticani
411. GIOVANNI DA UDINE [and assistants] *Grotesque Decorations* (detail), Rome, Vatican, Logge. Photo Anderson
412. GIOVANNI DA UDINE [and assistants] *Grotesque Decorations* (detail), Rome, Vatican, Logge. Photo Alinari
413. GIOVANNI DA UDINE *Borders from the Tapestries of the Acts of the Apostles*, Rome, Vatican Museum. Photo Alinari
414. POLIDORO *Grotesque Decorations*, Rome, Vatican, Stanza dell'Incendio. Photo Anderson
415. RAPHAEL [and assistants] *Logge* (first bay). Photo Alinari
416. RAPHAEL [and assistants] *Logge* (second bay). Photo Alinari
417. RAPHAEL [and assistants] *Logge* (third bay). Photo Alinari
418. RAPHAEL [and assistants] *Logge* (fourth bay). Photo Alinari
419. RAPHAEL [and assistants] *Logge* (fifth bay). Photo Alinari
420. RAPHAEL [and assistants] *Logge* (sixth bay). Photo Alinari
421. RAPHAEL [and assistants] *Logge* (seventh bay). Photo Alinari
422. GIOVANNI DA UDINE *Lower Loggia*, Rome, Vatican, Cortile di S. Damaso. Photo Gabinetto Fotografico Nazionale, Rome
423. RAPHAEL *Baldassare Castiglione*, Paris, Louvre. Photo Alinari
424. RAPHAEL *Antonio Tebaldeo* (from the *Parnassus* fresco), Rome, Vatican. Photo Archivio Fotografico delle Gallerie e Musei Vaticani
425. RAPHAEL *Antonio Tebaldeo* (copy), Florence, Uffizi. Photo Alinari
426. RAPHAEL *Andrea Navagero and Agostino Beazzano*, Rome, Galleria Doria. Photo Alinari
427. GIULIO *Bindo Altoviti*, Washington, National Gallery (Samuel H. Kress Collection). Photo museum
428. RAPHAEL *Cardinal Bernardo Bibbiena* (copy), Florence, Uffizi. Photo Alinari
429. RAPHAEL *Leo X with the Cardinals Giulio de' Medici and Luigi Rossi*, Florence, Pitti. Photo Sopr. alle Gallerie, Florence
430. RAPHAEL *Raphael and His Fencing Master*, Paris Louvre. Photo Alinari
431. GIULIO *Giovanna d'Aragona*, Paris, Louvre. Photo Alinari
432. RAPHAEL *Madonna della Tenda*, Munich, Pinakothek. Photo museum
433. RAPHAEL [with Penni] *Spasimo di Sicilia*, Madrid, Prado. Photo Anderson
434. RAPHAEL *Spasimo di Sicilia* (detail). Photo MAS, Barcelona
435. GIULIO *Cartoon for the Holy Family of Francis I* (fragment), Melbourne, National Gallery. Photo University Visual Aids Department. Courtesy of the Trustees of the National Gallery of Victoria, Melbourne

436. RAPHAEL [with Giulio]   *Holy Family of Francis I*, Paris, Louvre. Photo Alinari
437. RAPHAEL [with Giulio]   *St. Michael*, Paris, Louvre. Photo Alinari
438. RAPHAEL [and assistants]   *Transfiguration*, Rome, Vatican Museum. Photo Anderson
439. PENNI   *Transfiguration* (detail). Photo Anderson
440. RAPHAEL [and Giulio]   *Transfiguration* (detail). Photo Alinari
441. GIULIO   *Transfiguration* (detail). Photo Alinari
442. RAPHAEL   *Transfiguration* (detail), *St. Andrew*. Photo Alinari
443. RAPHAEL   *Study for St. Andrew*, London, British Museum. Photo museum. Courtesy of the Trustees of the British Museum
444. RAPHAEL   *Study for the Transfiguration*, Oxford, Ashmolean. Photo museum. Courtesy of the Visitors of the Ashmolean Museum
445. GIULIO (?)   *Study for the Transfiguration*, Paris, Louvre. Photo Alinari
446. GIULIO   *Madonna Piccola Gonzaga*, Paris, Louvre. Photo Archives Photographiques, Paris
447. GIULIO   *St. Margaret*, Paris, Louvre. Photo Alinari
448. GIULIO   *Madonna della Perla*, Madrid, Prado. Photo Anderson
449. PENNI   *Visitation*, Madrid, Prado. Photo Anderson
450. PENNI   *Madonna del Divino Amore*, Naples, Galleria Nazionale. Photo Brogi
451. GIULIO [with Penni]   *St. John Baptist*, Florence, Academy. Photo Sopr. alle Gallerie, Florence
452. GIULIO [with Raffaellino]   *Madonna della Rosa*, Madrid, Prado. Photo Anderson
453. GIULIO [with Raffaellino]   *Madonna of the Oak*, Madrid, Prado. Photo MAS, Barcelona
454. GIULIO [with Raffaellino]   *St. Margaret*, Vienna, Kunsthistorisches Museum. Photo Wolfrum
455. SEBASTIANO   *Man in Armor*, Hartford, Atheneum. Courtesy of Mr. H. Sperling, New York
456. SEBASTIANO   *Young Violinist*, Paris, Baron G. de Rothschild. Photo Archives Photographiques, Paris
457. SEBASTIANO   *Portrait of a Young Man*, Budapest, Museum. Photo museum
458. SEBASTIANO   *Cardinal Antonio Ciocchi del Monte Sansovino*, Dublin, National Gallery. Photo museum. Courtesy of the National Gallery of Ireland
459. SEBASTIANO   *Verdelotti and Ubretto* (destroyed; formerly Berlin, Museums). Photo museum
460. SEBASTIANO   *Cardinal Bandinello Sauli and Suite*, Washington, National Gallery, Kress Collection. Photo Samuel H. Kress Collection
461. SEBASTIANO   *Pietà*, Viterbo, Museum. Photo Gabinetto Fotografico Nazionale, Rome
462. SEBASTIANO   *Pietà*, Leningrad, Hermitage. Photo museum

463. SEBASTIANO *Resurrection of Lazarus*, London, National Gallery. Photo museum. Reproduced by courtesy of the Trustees, The National Gallery, London
464. SEBASTIANO *Resurrection of Lazarus* (detail). Photo Anderson
465. SEBASTIANO *Resurrection of Lazarus* (detail). Photo museum
466. SEBASTIANO *Holy Family with St. John Baptist and Donor*, London, National Gallery. Photo Anderson
467. SEBASTIANO *Cappella Borgherini*, Rome, S. Pietro in Montorio. Photo Fototeca di Architettura e Topografia dell'Italia Antica, Rome
468. SEBASTIANO *Two Prophets*, Cappella Borgherini. Photo Gabinetto Fotografico Nazionale, Rome
469. SEBASTIANO *Flagellation*, Cappella Borgherini. Photo Anderson
470. SEBASTIANO *Study for the Flagellation*, London, British Museum. Photo museum. Courtesy of the Trustees of the British Museum
471. MICHELANGELO *Modello for the Flagellation* (copy), Windsor, Royal Library. Photo Royal Library. Copyright reserved
472. SEBASTIANO *Study for the Flagellation*, London, British Museum. Photo museum. Courtesy of the Trustees of the British Museum
473. SEBASTIANO *Transfiguration*, Cappella Borgherini. Photo Gabinetto Fotografico Nazionale, Rome
474. SEBASTIANO *Visitation*, Paris, Louvre. Photo Alinari
475. SEBASTIANO *Martyrdom of St. Agatha*, Florence, Pitti. Photo Alinari
476. SEBASTIANO *Study for St. Agatha*, Paris, Louvre. Photo museum
477. VIRGILIO ROMANO *House in Vicolo del Campanile*, Rome. Photo Fototeca di Architettura e Topografia dell'Italia Antica, Rome
478. PERUZZI *Sala delle Prospettive*, Rome, Villa Farnesina. Photo Fototeca di Architettura e Topografia dell'Italia Antica, Rome
479. PERUZZI *Sala delle Prospettive*. Photo Fototeca di Architettura e Topografia dell'Italia Antica, Rome
480. PERUZZI *Sala delle Prospettive* (detail). Photo Fototeca di Architettura e Topografia dell'Italia Antica, Rome
481. PERUZZI *Death of Adonis*, Sala delle Prospettive. Photo Alinari
482. PERUZZI *Procession of Bacchus*, Sala delle Prospettive. Photo Alinari
483. PERUZZI *Ducalion and Pyrrha*, Sala delle Prospettive. Photo Alinari
484. PERUZZI *Venus and Cupid*, Sala delle Prospettive. Photo Alinari
485. PERUZZI *Apollo*, Sala delle Prospettive. Photo Fototeca di Architettura e Topografia dell'Italia Antica, Rome
486. PERUZZI (?) [and Marcantonio] *Quos Ego*. Photo Sopr. alle Gallerie, Florence
487. PERUZZI [and Ugo da Carpi] *Hercules Expelling Envy from the Temple of the Muses*. Photo Sopr. alle Gallerie, Florence
488. PERUZZI *Apollo and the Muses*, Florence, Pitti. Photo Sopr. alle Gallerie, Florence

489. PERUZZI *Country Festival*, Florence, Uffizi. Photo Sopr. alle Gallerie, Florence
490. PERUZZI *Ponzetti Chapel* (vault frescoes), Rome, S. M. della Pace. Photo Anderson
491. PERUZZI *Ponzetti Chapel* (vault frescoes, detail). Photo Anderson
492. PERUZZI *Ponzetti Chapel* (altar fresco). Photo Anderson
493. PERUZZI *Portrait of a Carmelite*, Rome, Art Market. Photo courtesy of Signore Alessandro Morandotti
494. PERUZZI *Presentation of the Virgin*, Rome, S.M. della Pace. Photo Anderson
495. PERUZZI [and assistants] *Decoration in the Palazzo della Cancelleria*, Rome. Photo Alinari
496. PERUZZI [and assistants] *Decoration in the Palazzo della Cancelleria*, Rome. Photo Alinari
497. PERUZZI *Joseph put into the Well*, Cancelleria. Photo Archivio Fotografico delle Gallerie e Musei Vaticani
498. PERUZZI *Meeting of Solomon and Sheba*, Cancelleria. Photo Archivio Fotografico delle Gallerie e Musei Vaticani
499. PENNI (?) *Separation of Light and Darkness*, Rome, Vatican, Logge. Photo Anderson
500. GIULIO *Expulsion*, Logge. Photo Gabinetto Fotografico Nazionale, Rome
501. GIULIO *God Appearing to Isaac*, Logge. Photo Anderson
502. GIULIO *Isaac and Rebecca*, Logge. Photo Anderson
503. GIULIO [with Polidoro?] *Jacob and Rachel*, Logge. Photo Gabinetto Fotografico Nazionale, Rome
504. GIULIO (?) [with Perino] *Flight of Jacob*, Logge. Photo Anderson
505. GIULIO (?) *Crossing of the Red Sea*, Logge. Photo Anderson
506. GIULIO (?) [with Polidoro] *Moses Striking Water from the Rock*, Logge. Photo Anderson
507. GIULIO *Moses Receiving the Tablets of the Law*, Logge. Photo Anderson
508. GIULIO (?) *Adoration of the Golden Calf*, Logge. Photo Anderson
509. PERINO *Fall of Jericho*, Logge. Photo Anderson
510. PERINO *Joshua Stays the Sun*, Logge. Photo Gabinetto Fotografico Nazionale, Rome
511. PERINO *Division of the Lands*, Logge. Photo Gabinetto Fotografico Nazionale, Rome
512. PERINO *David and Goliath*, Logge. Photo Anderson
513. PERINO *David and Bathsheba*, Logge. Photo Brogi
514. PELLEGRINO (?) *Judgment of Solomon*, Logge. Photo Gabinetto Fotografico Nazionale, Rome
515. POLIDORO *Meeting of Solomon and Sheba*, Logge. Photo Gabinetto Fotografico Nazionale, Rome
516. POLIDORO *Building of the Temple*, Logge. Photo Anderson

517. PERINO *Adoration of the Shepherds*, Logge. Photo Anderson
518. PERINO *Adoration of the Kings*, Logge. Photo Brogi
519. PERINO *Baptism of Christ*, Logge. Photo Chauffourier, Rome
520. PERINO *Last Supper*, Logge. Photo Gabinetto Fotografico Nazionale, Rome
521. PENNI *Study for Separation of Light and Darkness*, London, British Museum. Photo museum. Courtesy of the Trustees of the British Museum
522. PENNI *Study for Expulsion*, Windsor, Royal Library. Photo Royal Library. Copyright reserved
523. PENNI *Study for Jacob's Dream*, London, British Museum. Photo museum. Courtesy of the Trustees of the British Museum
524. PENNI *Study for Adoration of the Golden Calf*, Florence, Uffizi. Photo Sopr. alle Gallerie, Florence
525. PERINO *Study for Division of the Lands*, Windsor, Royal Library. Photo Royal Library. Copyright reserved
526. PERINO *Lamentation over Christ*, Paris, Louvre. Photo Museum
527. PERINO *Pietà*, Rome, Sto. Stefano del Cacco. Photo Fototeca di Architettura e Topografia dell'Italia Antica, Rome
528. PERINO *Adoration of the Child*, Rome, Borghese 464. Photo Anderson

VI

CLIMAX AND CRISIS IN FLORENCE AND THE GENERATION OF FLORENTINE MANNERISM (*c*. 1514–1520)

529. FRA BARTOLOMMEO *Virgin and Child*, Florence, Convent of S. Marco. Photo Alinari
530. FRA BARTOLOMMEO *St. Mark Evangelist*, Florence, Pitti. Photo Sopr. alle Gallerie, Florence
531. FRA BARTOLOMMEO *Study for the Madonna della Misericordia*, Florence, Uffizi. Photo Sopr. alle Gallerie, Florence
532. FRA BARTOLOMMEO *Madonna della Misericordia*, Lucca, Pinacoteca. Photo Brogi
533. FRA BARTOLOMMEO *Annunciation*, Le Caldine, Convento della Maddalena. Photo Sopr. alle Gallerie, Florence
534. FRA BARTOLOMMEO *Study for the Caldine Annunciation*, Florence, Uffizi. Photo Sopr. alle Gallerie, Florence
535. FRA BARTOLOMMEO *Annunciation Altar*, Paris, Louvre. Photo Alinari
536. FRA BARTOLOMMEO *Salvator Mundi*, Florence, Pitti. Photo Alinari
537. FRA BARTOLOMMEO *Studies for the Salvator Mundi and Other Projects*, Florence, Uffizi. Photo Sopr. alle Gallerie, Florence
538. FRA BARTOLOMMEO *Study for Salvator Mundi*, Amsterdam, Rijksmuseum. Photo museum
539. FRA BARTOLOMMEO *Presentation in the Temple*, Vienna, Kunsthistorisches Museum. Photo museum

540. FRA BARTOLOMMEO *Isaiah*, Florence, Academy. Photo Anderson
541. FRA BARTOLOMMEO *Job*, Florence, Academy. Photo Anderson
542. FRA BARTOLOMMEO [with Fra Paolino] *Assumption of the Virgin*, Naples, Galleria Nazionale. Photo Alinari
543. FRA BARTOLOMMEO *Study for the Assumption*, Munich, Graphische Sammlung. Photo museum
544. FRA BARTOLOMMEO *Holy Family*, Rome, Galleria Nazionale. Photo Alinari
545. FRA BARTOLOMMEO *Madonna and Child with Elizabeth and John*, London, Kenwood, Iveagh Bequest. Photo courtesy of the Trustees of the Cook Collection
546. FRA BARTOLOMMEO [with Bugiardini] *Pietà*, Florence, Pitti. Photo Alinari
547. FRA BARTOLOMMEO *Noli Me Tangere*, Le Caldine, Convento della Maddalena. Photo Alinari
548. FRA PAOLINO *Crucifixion*, Siena, Sto. Spirito. Photo Alinari
549. FRA PAOLINO *Holy Family with St. John and Angels*, Rome, Galleria Doria. Photo Gabinetto Fotografico Nazionale, Rome
550. FRA PAOLINO *Pietà*, Florence, Museo di S. Marco. Photo Alinari
551. ANDREA DEL SARTO *Scenes from the Life of St. John Baptist*, Florence, Scalzo. Photo author
552. ANDREA DEL SARTO *Preaching of St. John Baptist*, Florence, Scalzo. Photo Brogi
553. ANDREA DEL SARTO *Justice*, Florence, Scalzo. Photo Alinari
554. ANDREA DEL SARTO *Charity*, Florence, Scalzo. Photo Brogi
555. ANDREA DEL SARTO *Chiaroscuro Panel* (from the Leo X Festival Decoration?), Florence, Uffizi. Photo Sopr. alle Gallerie, Florence
556. ANDREA DEL SARTO *Chiaroscuro Panel* (from the Leo X Festival Decoration?), Florence, Uffizi. Photo Sopr. alle Gallerie, Florence
557. ANDREA DEL SARTO *Chiaroscuro Panel* (from the Leo X Festival Decoration?), Florence, Uffizi. Photo Sopr. alle Gallerie, Florence
558. ANDREA DEL SARTO *Holy Family with St. Catherine*, Leningrad, Hermitage. Photo museum
559. A. VENEZIANO [after Andrea del Sarto] *Pietà*. Photo courtesy of Metropolitan Museum of Art, Whittelsey Fund
560. ANDREA DEL SARTO *Study for the Pietà*, Florence, Uffizi. Photo Alinari
561. ANDREA DEL SARTO *Redeemer*, Florence, SS. Annunziata. Photo Alinari
562. ANDREA DEL SARTO *Holy Family*, Paris, Louvre 1515. Photo Alinari
563. ANDREA DEL SARTO *Baptism of the Multitude*, Florence, Scalzo. Photo Brogi
564. ANDREA DEL SARTO *Capture of St. John*, Florence, Scalzo. Photo Brogi
565. ANDREA DEL SARTO *Madonna of the Harpies*, Florence, Uffizi. Photo Brogi

566. ANDREA DEL SARTO *Disputation on the Trinity*, Florence, Pitti. Photo Anderson
567. ANDREA DEL SARTO *Portrait of a Sculptor*(?), London, National Gallery. Photo museum. Reproduced by courtesy of the Trustees, The National Gallery, London
568. ANDREA DEL SARTO *Study for Portrait of a Sculptor*, Florence, Uffizi. Photo Sopr. alle Gallerie, Florence
569. ANDREA DEL SARTO *Caritas*, Paris, Louvre. Photo Alinari
570. ANDREA DEL SARTO *Madonna and Child with St. John*, Rome, Borghese 334. Photo Anderson
571. ANDREA DEL SARTO *Madonna with St. John and Angels* (?), London, Wallace Collection. Photo reproduced by permission of the Trustees of the Wallace Collection
572. ANDREA DEL SARTO *Pietà*, Vienna, Kunsthistorisches Museum. Photo Wolfrum
573. ANDREA DEL SARTO *Tribute to Caesar* (in its original dimensions), Poggio a Cajano. Photo Alinari
574. FRANCIABIGIO *Annunciation*, Turin, Galleria Sabauda. Photo Alinari
575. FRANCIABIGIO *Angel*, Florence, Sto. Spirito. Photo Brogi
576. FRANCIABIGIO *St. Job Altar*, Florence, Uffizi. Photo Brogi
577. FRANCIABIGIO *Meeting of Christ and St. John Baptist*, Florence, Scalzo. Photo Alinari
578. FRANCIABIGIO *Madonna and Child with St. John*, Vienna, Kunsthistorisches Museum 208. Photo museum
579. FRANCIABIGIO *Benediction of St. John by Zachary*, Florence, Scalzo. Photo Alinari
580. FRANCIABIGIO *Leda*, Brussels, Museum. Photo copyright A.C.L. Brussels
581. FRANCIABIGIO *Madonna*, Bologna, Pinacoteca 294. Photo Anderson
582. FRANCIABIGIO *Triumph of Caesar* (in its original dimensions), Poggio a Cajano. Photo Alinari
583. FRANCIABIGIO *Self-Portrait*, New York, Hunter College, ex Kress Collection. Photo Samuel H. Kress Collection, New York
584. FRANCIABIGIO *Portrait of Caradosso*, London, Art Market. Photo A. C. Cooper
585. FRANCIABIGIO *Portrait of a Man*, Vienna, Liechtenstein Collection. Photo Wolfrum
586. FRANCIABIGIO *Portrait of a Fattore*, Hampton Court. Photo Hampton Court. Copyright reserved
587. BUGIARDINI *Temptation in the Garden of Eden*, New York, Private Collection. Photo courtesy of Duveen Brothers, New York
588. BUGIARDINI *St. Sebastian*, New York, Kress Collection. Photo Samuel H. Kress Collection, New York

589. BUGIARDINI *Madonna and Child with the Infant St. John*, Dunblane, Stirling Collection. Photo Talani, Florence
590. BUGIARDINI *Madonna and Child with the Infant St. John*, Florence, Uffizi. Photo Anderson
591. BUGIARDINI *Madonna and Child with the Infant St. John*, Allentown (Pennsylvania), Museum, Kress Collection. Photo Samuel H. Kress Collection, New York
592. RIDOLFO GHIRLANDAIO *Coronation of the Virgin*, Florence, S.M. Novella, Cappella del Papa. Photo Brogi
593. RIDOLFO GHIRLANDAIO *Resuscitation of a Youth by St. Zenobius*, Florence, Academy. Photo Anderson
594. RIDOLFO GHIRLANDAIO *Translation of the Body of St. Zenobius*, Florence, Academy. Photo Anderson
595. RIDOLFO GHIRLANDAIO *Madonna with Six Saints*, Pistoia, Museum. Photo Brogi
596. RIDOLFO GHIRLANDAIO *Girolamo Benivieni* (?), London, National Gallery. Photo Anderson
597. RIDOLFO GHIRLANDAIO *Portrait of a Man*, Florence, Galleria Corsini. Photo Alinari
598. RIDOLFO GHIRLANDAIO *Portrait of a Man*, Florence, Torrigiani Collection. Photo Talani, Florence
599. RIDOLFO GHIRLANDAIO *Pietà*, Colle di Val d'Elsa, S. Agostino. Photo Brogi
600. GRANACCI *Madonna with Sts. Francis and Zenobius*, Florence, Academy. Photo Alinari
601. GRANACCI *Holy Family*, Boughton House, Duke of Buccleuch. Photo Ideal Studios, Edinburgh. Courtesy of the Duke of Buccleuch
602. GRANACCI *Madonna and Child*, San Francisco, Palace of the Legion of Honor. Photo museum
603. GRANACCI *Sts. John, Apollonia, Mary Magdalen, and Jerome* (from the *Apollonia Altar*), Munich, Pinakothek. Photo museum
604. GRANACCI *Predella Panel* (from the *Apollonia Altar*), Florence, Academy. Photo Sopr. alle Gallerie, Florence
605. GRANACCI *Predella Panel* (from the *Apollonia Altar*), Florence, Academy. Photo Sopr. alle Gallerie, Florence
606. GRANACCI *Predella Panel* (from the *Apollonia Altar*), Florence, Academy. Photo Sopr. alle Gallerie, Florence
607. GRANACCI *Entry of Charles VIII into Florence*, Florence, Museo Mediceo. Photo Anderson
608. GRANACCI *Joseph Presents His Father to Pharaoh*, Florence, Uffizi. Photo Alinari
609. GRANACCI *Arrest of Joseph*, Florence, Uffizi. Photo Alinari
610. GRANACCI *Madonna with St. John*, (formerly) Munich, A. S. Drey. Photo courtesy of Drey Galleries, New York

611. GRANACCI *Madonna with Four Saints*, Montemurlo, Pieve. Photo Sopr. alle Gallerie, Florence
612. SOGLIANI *Madonna with St. John*, Baltimore, Walters Art Gallery. Photo courtesy of Walters Art Gallery
613. SOGLIANI *Madonna with St. John*, Brussels, Museum. Photo copyright A.C.L. Brussels
614. SOGLIANI *Madonna with St. John*, Turin, Galleria Sabauda. Photo Alinari
615. SOGLIANI *S. Acasio Altar*, Florence, S. Lorenzo. Photo Alinari
616. SOGLIANI *S. Brigitta Altar*, Florence, Academy. Photo Alinari
617. PULIGO [on Sarto's cartoon] *Holy Family*, London, National Gallery. Photo museum. Reproduced by courtesy of the Trustees, The National Gallery, London
618. ANDREA [with Puligo] *Story of Joseph* (1), Florence, Pitti 87. Photo Brogi
619. ANDREA [with Puligo] *Story of Joseph* (2), Florence, Pitti 88. Photo Sopr. alle Gallerie, Florence
620. PULIGO *Madonna with St. John*, Florence, Pitti 242. Photo Alinari
621. PULIGO *Madonna with St. John and Angels*, Florence, Galleria Corsini. Photo Brogi
622. PULIGO *Adoration of the Kings* (formerly) Milan, Crespi Collection. Photo Anderson
623. PULIGO *Deposition*, Venice, Seminario. Photo Anderson
624. PULIGO *Preaching of St. John Baptist* (formerly) London, Henry Harris Collection. Photo A. C. Cooper
625. PULIGO *History of Joseph*, Rome, Borghese 463. Photo Sopr. alle Gallerie, Florence
626. PULIGO *Apollo and Daphne*, Florence, Galleria Corsini. Photo Alinari
627. BACCHIACCA *Deposition*, Bassano, Museo Civico. Photo Alinari
628. BACCHIACCA *Adam and Eve*, Philadelphia Museum, Johnson Collection. Photo courtesy of John G. Johnson Collection, Philadelphia
629. BACCHIACCA *Story of Joseph*, London, National Gallery 1218. Photo museum. Reproduced by courtesy of the Trustees, The National Gallery, London
630. BACCHIACCA *Story of Joseph*, Rome, Borghese. Photo Sopr. alle Gallerie, Florence
631. BACCHIACCA *Story of Joseph*, London, National Gallery 1219. Photo museum. Reproduced by courtesy of the Trustees, The National Gallery, London
632. BACCHIACCA *Story of Joseph*, Rome, Borghese. Photo Sopr. alle Gallerie, Florence
633. BACCHIACCA *Deposition*, Florence, Uffizi. Photo Brogi
634. BACCHIACCA *Creation of Eve*, Stockholm, Private Collection. Photo Talani, Florence

635. BACCHIACCA *Leda*, Rotterdam, Boymans Museum, van Beuningen Collection. Photo museum
636. BACCHIACCA *Predella Panel* (from the *S. Acasio Altar*), Florence, Uffizi. Photo Anderson
637. BACCHIACCA *Predella Panel* (from the *S. Acasio Altar*), Florence, Uffizi. Photo Anderson
638. BACCHIACCA *Legend of the Dead King*, Dresden, Gallery. Photo Alinari
639. PONTORMO *Visitation*, Florence, SS. Annunziata. Photo Sopr. alle Gallerie, Florence
640. PONTORMO *Visitation* (detail). Photo Sopr. alle Gallerie, Florence
641. PONTORMO *Cappella del Papa* (vault), Florence, S.M. Novella. Photo Brogi
642. PONTORMO *St. Veronica*, Florence, S.M. Novella, Cappella del Papa. Photo Sopr. alle Gallerie, Florence
643. PONTORMO *Joseph Revealing Himself to His Brothers*, Henfield, Lady Salmond. Photo Sopr. alle Gallerie, Florence
644. PONTORMO *Joseph Sold to Potiphar*, Henfield, Lady Salmond. Photo Sopr. alle Gallerie, Florence
645. PONTORMO *The Butler Restored and the Baker Led to Execution*, Henfield, Lady Salmond. Photo Sopr. alle Gallerie, Florence
646. PONTORMO *Madonna and Saints*, Florence, S. Michele Visdomini. Photo Sopr. alle Gallerie, Florence
647. PONTORMO *Pietà* (predella for the Visdomini Altar), Dublin, National Gallery. Photo museum. Courtesy of the National Gallery of Ireland
648. PONTORMO *St. Lawrence* (portion of predella), Dublin, National Gallery. Photo museum. Courtesy of the National Gallery of Ireland
649. PONTORMO *St. Francis* (portion of predella), Dublin, National Gallery. Photo museum. Courtesy of the National Gallery of Ireland
650. PONTORMO *Joseph in Egypt*, London, National Gallery. Photo museum. Reproduced by courtesy of the Trustees, The National Gallery, London
651. PONTORMO *Cosimo de' Medici*, Florence, Uffizi. Photo Sopr. alle Gallerie, Florence
652. PONTORMO *Study for a Portrait of Piero de' Medici*, Rome, Galleria Corsini. Photo Sopr. alle Gallerie, Florence
653. PONTORMO *Portrait of a Musician* [Francesco dell'Ajolle?], Florence, Uffizi. Photo Sopr. alle Gallerie, Florence
654. PONTORMO *Study for St. John Evangelist*, Florence, Uffizi. Photo Sopr. alle Gallerie, Florence
655. PONTORMO *St. John Evangelist*, Empoli, Collegiata. Photo Sopr. alle Gallerie, Florence
656. PONTORMO *St. Michael*, Empoli, Collegiata. Photo Sopr. alle Gallerie, Florence
657. PONTORMO *Study for a Pietà*, Florence, Uffizi. Photo Sopr. alle Gallerie, Florence

658. PONTORMO *St. Anthony Abbot*, Florence, Uffizi. Photo Sopr. alle Gallerie, Florence
659. BERRUGUETE *Salome*, Florence, Uffizi. Photo Sopr. alle Gallerie, Florence
660. BERRUGUETE *Madonna*, Florence, Uffizi. Photo Sopr. alle Gallerie, Florence
661. BERRUGUETE *Madonna with St. John*, Florence, Palazzo Vecchio, Loeser Collection. Photo Brogi
662. ROSSO *Madonna in Glory*, Leningrad, Hermitage. Photo Sopr. alle Gallerie, Florence
663. ROSSO *Portrait of a Young Man*, Berlin, Museums. Photo museum
664. ROSSO *Assumption*, Florence, SS. Annunziata. Photo Sopr. alle Gallerie, Florence
665. ROSSO *Assumption* (detail). Photo Sopr. alle Gallerie, Florence
666. ROSSO *Memento Mori*, Florence, Uffizi. Photo Sopr. alle Gallerie, Florence
667. ROSSO *Madonna and Saints (S.M. Nuova Altar)*, Florence, Uffizi. Photo Alinari
668. ROSSO *S.M. Nuova Altar* (detail). Photo Alinari

VII

EPILOGUE: THE ASCENDANCY OF MANNERISM
(*some events of 1521*)

669. ROSSO *Madonna with Sts. John Baptist and Bartholomew*, Villamagna (near Volterra), Pieve. Photo Sopr. alle Gallerie, Florence
670. ROSSO *Deposition*, Volterra, Museum. Photo Sopr. alle Gallerie, Florence
671. ROSSO *Deposition* (detail). Photo Sopr. alle Gallerie, Florence
672. ROSSO *Deposition* (detail). Photo Sopr. alle Gallerie, Florence
673. ROSSO *Deposition* (detail). Photo Sopr. alle Gallerie, Florence
674. PONTORMO *Design for Wall at Poggio a Cajano*, London, British Museum. Photo Alinari
675. PONTORMO *First Project for Lunette at Poggio*, Florence, Uffizi. Photo Sopr. alle Gallerie, Florence
676. PONTORMO *Second Project for Lunette at Poggio*, Florence, Uffizi. Photo Sopr. alle Gallerie, Florence
677. PONTORMO *Study for Lunette at Poggio*, Florence, Uffizi. Photo Sopr. alle Gallerie, Florence
678. PONTORMO *Vertumnus and Pomona*, Poggio a Cajano. Photo Alinari
679. PONTORMO *Vertumnus and Pomona*, Poggio a Cajano. Photo Alinari
680. PONTORMO *Vertumnus and Pomona* (detail). Photo Sopr. alle Gallerie, Florence
681. GIOVANNI DA UDINE [and assistants] *Loggia*, Rome, Villa Madama. Photo Anderson

682. GIOVANNI DA UDINE [with Peruzzi]   *Loggia* (central vault), Villa Madama. Photo Anderson
683. GIOVANNI DA UDINE [with Peruzzi]   *Loggia* (side vault), Villa Madama. Photo Anderson
684. GIOVANNI DA UDINE [with Peruzzi]   *Loggia* (side vault), Villa Madama. Photo Anderson
685. PERUZZI   *Loggia* (detail), Villa Madama. Photo Alinari
686. PERUZZI   *Loggia* (detail), Villa Madama. Photo Anderson
687. GIOVANNI DA UDINE [with Perino]   *Sala dei Pontefici*, Rome, Vatican. Photo Anderson
688. GIOVANNI DA UDINE [with Perino]   *Sala dei Pontefici* (detail). Photo Anderson
689. PERINO   *Sala dei Pontefici* (detail). Photo Brogi
690. PERINO   *Sala dei Pontefici* (detail). Photo Brogi
691. PERINO   *Sala dei Pontefici* (central panel of the vault). Photo Anderson
692. GIULIO   *Justice* (trial figure), Rome, Vatican, Sala di Costantino, Rome, Vatican. Photo Anderson
693. GIULIO [and assistants]   *Sala di Costantino*. Photo Archivio Fotografico delle Gallerie e Musei Vaticani
694. GIULIO [and assistants]   *Sala di Costantino*. Photo Alinari
695. GIULIO   *Allocutio*, Sala di Costantino. Photo Anderson
696. GIULIO [and assistants]   *The Battle of Constantine*, Sala di Costantino. Photo Anderson
697. GIULIO   *St. Peter with Ecclesia and Aeternitas*, Sala di Costantino. Photo Anderson
698. GIULIO   *Leo X as Clement I with Moderatio and Comitas*, Sala di Costantino. Photo Anderson
699. GIULIO [and assistants]   *Battle of Constantine* (detail), Sala di Costantino. Photo Anderson
700. POLIDORO   *Basamento*, Sala di Costantino. Photo Archivio Fotografico delle Gallerie e Musei Vaticani

I

INTRODUCTION

THE GENESIS OF HIGH RENAISSANCE CLASSICAL STYLE

1. LEONARDO *Head of an Angel*, from Verrocchio's *Baptism of Christ*, Florence, Uffizi

2. VERROCCHIO *Baptism of Christ*, Florence, Uffizi

3. LEONARDO *Adoration of the Magi*, Florence, Uffizi

4. LEONARDO *Adoration of the Magi* (detail)

5. LEONARDO *Adoration of the Magi* (detail)

6. DOMENICO GHIRLANDAIO
Adoration of the Magi,
Florence, Innocenti

7. ANTONIO POLLAIUOLO *Hercules and the Hydra, Hercules and Anteus* (lost; formerly Florence, Uffizi)

8. PIERO DELLA FRANCESCA
Montefeltro Altar, Milan, Brera

9. BOTTICELLI *Adoration of the Magi*, Florence, Uffizi

10. LEONARDO *Virgin of the Rocks*, Paris, Louvre

11. LEONARDO *Virgin of the Rocks* (detail)

12. LEONARDO *Last Supper*, Milan, S.M. delle Grazie

13. LEONARDO *Last Supper* (detail)

14. FILIPPINO LIPPI *Adoration of the Magi*, Florence, Uffizi

15. PIERO DI COSIMO *Mars and Venus*, Berlin, Museums

16. PERUGINO *Vision of St. Bernard*, Munich, Pinakothek

17. MICHELANGELO
Madonna of the Steps, Florence,
Casa Buonarroti

18. MICHELANGELO *Battle of the Centaurs*, Florence, Casa Buonarroti

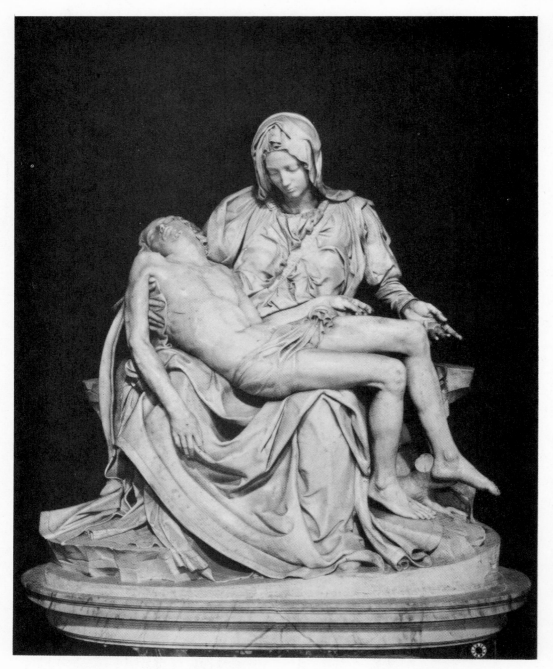

19. MICHELANGELO *Pietà*, Rome, St. Peter's

20. MICHELANGELO *Bacchus*, Florence, Bargello

21. FRA BARTOLOMMEO [with Albertinelli] *Last Judgment*, Florence, Museo di S. Marco

II
FORMATION OF THE CLASSICAL VOCABULARY
(c. 1500-c. 1508)

22. LEONARDO *St. Anne Cartoon*, London, Royal Academy

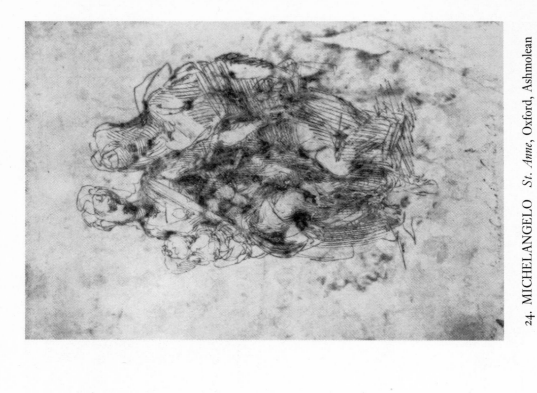

24. MICHELANGELO *St. Anne*, Oxford, Ashmolean

23. ANDREA DEL BRESCIANINO *St. Anne* (destroyed; formerly Berlin, Museums)

25. MICHELANGELO *Bruges Madonna*, Bruges, Notre Dame

26. MICHELANGELO *David*, Florence, Academy

27. MICHELANGELO *Doni Holy Family*, Florence, Uffizi

28. LEONARDO *Battle of the Standard* (copy), Florence, Palazzo Vecchio

29. RUBENS *Battle of the Standard* (copy after Leonardo), Paris, Louvre

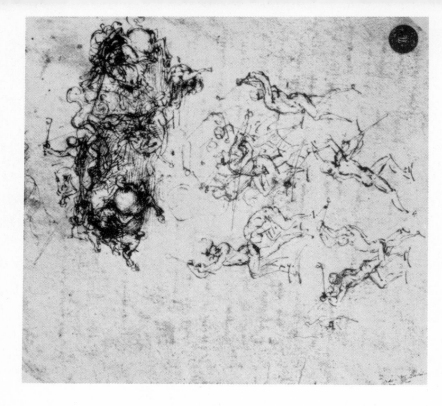
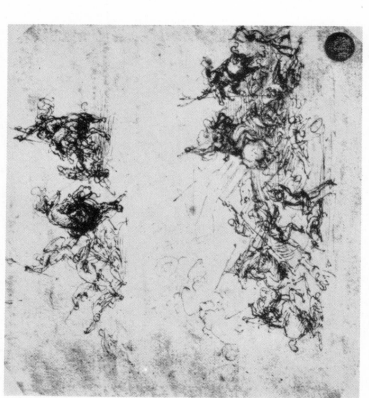

30, 31. LEONARDO *Studies for the Battle of Anghiari*, Venice, Academy

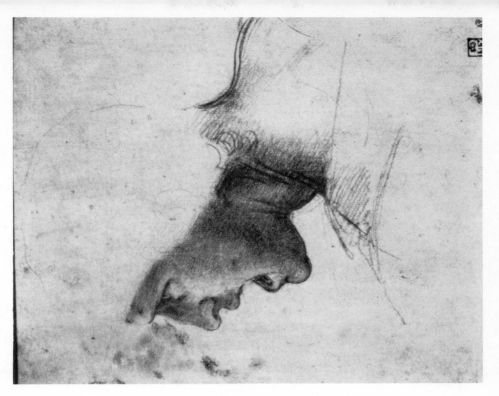

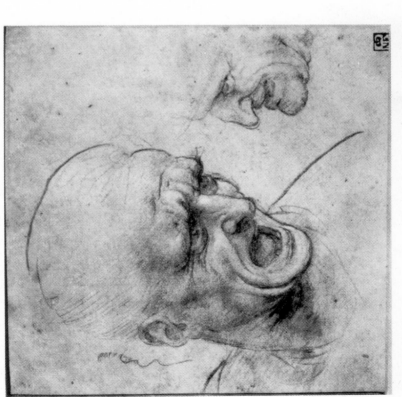

32, 33. LEONARDO *Studies for the Battle of Anghiari*, Budapest, Museum

34. MICHELANGELO
St. Matthew, Florence, Academy

35. MICHELANGELO *Study for the Julius Tomb* (incorporating the lower story of the project of 1505[?]; copy), Florence, Uffizi

36. MICHELANGELO *Battle of Cascina* (central portion); copy attributed to Aristotile da Sangallo, Holkham Hall

37. LEONARDO *Mona Lisa*, Paris, Louvre

38. LEONARDO *Mona Lisa* (detail)

39. LEONARDO *St. Anne, Virgin, and Child*, Paris, Louvre

40. ALBERTINELLI *Madonna and Saints* (portable triptych), Milan, Poldi-Pezzoli

41. ALBERTINELLI *Crucifixion*, Florence, Certosa

42. ALBERTINELLI *Visitation*, Florence, Uffizi

43. FRA BARTOLOMMEO *Vision of St. Bernard*, Florence, Academy

44. FILIPPINO LIPPI *Vision of St. Bernard*, Florence, Badia

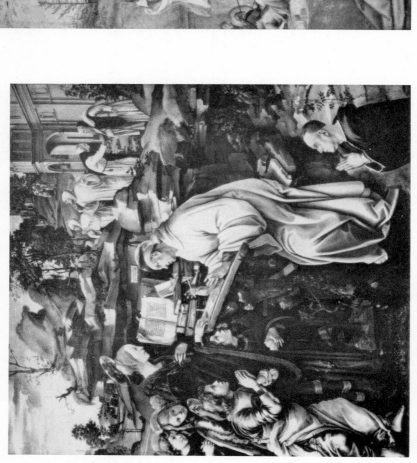

45. FRA BARTOLOMMEO *Noli Me Tangere*, Paris, Louvre

46. FRA BARTOLOMMEO [with Albertinelli] *Assumption of the Virgin* (destroyed; formerly Berlin, Museums)

47. ALBERTINELLI *Annunciation with Sts. Sebastian and Lucy*, Munich, Pinakothek

48. RAPHAEL *Three Graces*, Chantilly, Musée Condé

49. RAPHAEL   *Marriage of the Virgin*, Milan, Brera

50. RAPHAEL *Madonna del Granduca*, Florence, Pitti

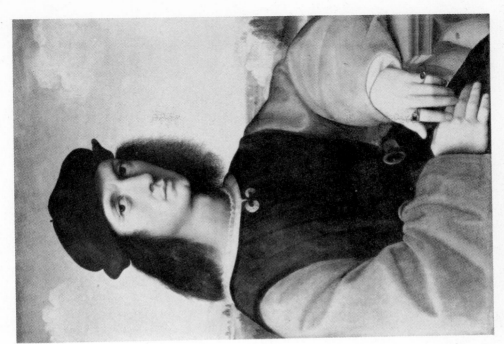

52. RAPHAEL *Angelo Doni*, Florence, Pitti

51. RAPHAEL *Maddalena Doni*, Florence, Pitti

53. RAPHAEL *Madonna del Prato*, Vienna, Kunsthistorisches Museum

54. RAPHAEL *Madonna del Cardellino*, Florence, Uffizi

55. RAPHAEL *La Belle Jardinière*, Paris, Louvre

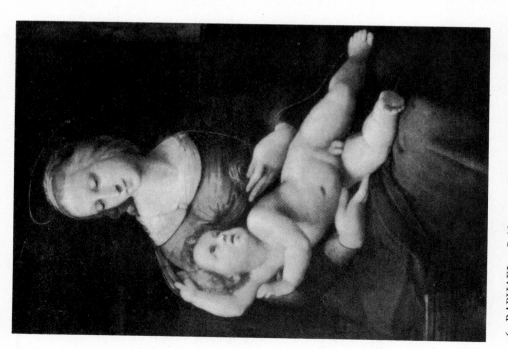

56. RAPHAEL *Bridgewater House Madonna*, Edinburgh, National Gallery, Ellesmere Loan

57. RAPHAEL *Madonna Study*, Paris, Louvre

58. RAPHAEL *Holy Family of the Casa Canigiani*, Munich, Pinakothek

59. RAPHAEL *Entombment*, Rome, Borghese

60. RAPHAEL *Madonna del Baldacchino*, Florence, Pitti

61. RAFFAELLINO CARLI *Madonna with Saints*, Florence, Sto. Spirito

62. LORENZO DI CREDI *Madonna with Saints*, Pistoia, S.M. delle Grazie

63. PIERO DI COSIMO *Madonna with Saints*, Florence, Innocenti

64. PIERO DI COSIMO *Immaculate Conception*, Florence, Uffizi

65. PIERO DI COSIMO *Madonna with St. John*, Vienna, Liechtenstein Collection

66. PIERO DI COSIMO *Liberation of Andromeda*, Florence, Uffizi 1536

67. GRANACCI *Holy Family*, Washington, National Gallery, Kress Collection

68. GRANACCI *Madonna with Two Saints*, Berlin, Museums

69. GRANACCI *Holy Family with St. John*, Dublin, National Gallery

71. BUGIARDINI *Madonna and Child with St. John*, New York, Metropolitan Museum
FRA BARTOLOMMEO

70. GRANACCI *Madonna della Cintola*, Florence, Academy

73. BUGIARDINI *Portrait of a Lady*, Urbino, Galleria Nazionale
RAPHAEL (?)

72. BUGIARDINI *Holy Family*, Turin, Galleria Sabauda

74. RIDOLFO GHIRLANDAIO *Madonna with Sts. Francis and Mary Magdalen*, Florence, Academy

75. RIDOLFO GHIRLANDAIO *Coronation of the Virgin*, Paris, Louvre

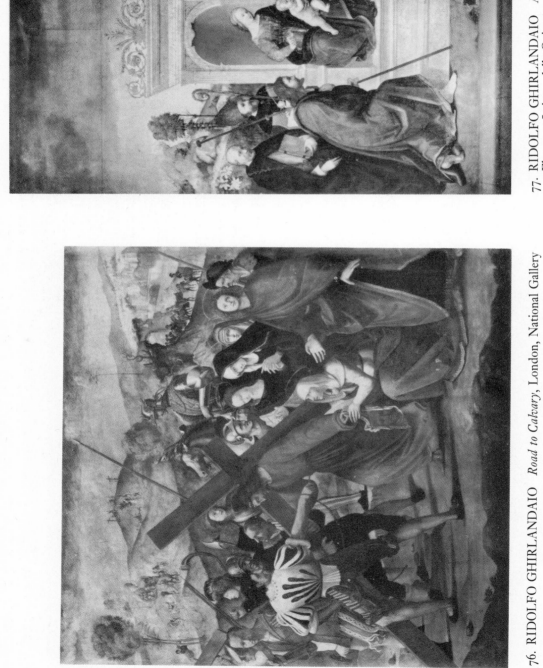

76. RIDOLFO GHIRLANDAIO *Road to Calvary*, London, National Gallery

77. RIDOLFO GHIRLANDAIO *Marriage of St. Catherine*, Florence, Istituto delle Quiete

78, 79. RIDOLFO GHIRLANDAIO   *Altar Wings with Angels*, Florence, Academy

81. RIDOLFO GHIRLANDAIO *Lady with a Rabbit*, New Haven, Yale University Art Gallery

80. RIDOLFO GHIRLANDAIO *Portrait of a Man*, Chicago, Art Institute

82. RIDOLFO GHIRLANDAIO *Girl with a Unicorn*, Rome, Borghese

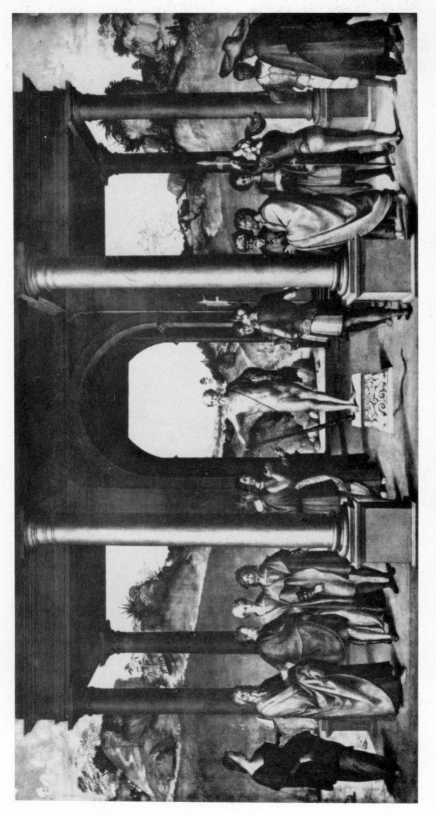

83. FRANCIABIGIO *Temple of Hercules*, Florence, Palazzo Davanzati

85. FRANCIABIGIO *Madonna*, Perugia, Count Ranieri

84. FRANCIABIGIO *Madonna and Child with St. John*, Florence, Uffizi 2178

87. FRANCIABIGIO *Madonna del Pozzo*, Florence, Academy

86. FRANCIABIGIO *Holy Family*, Florence, Academy (ex-Uffizi 888)

88. FRANCIABIGIO *Holy Family*, Vienna, Kunsthistorisches Museum 206

89. PINTURICCHIO   *Choir Vault*, Rome, S.M. del Popolo

90. [Anon.]   *Sibyls*, Rome, S. Pietro in Montorio

91. PERUZZI [and others] *Choir Decoration*, Rome, S. Onofrio

92. PERUZZI (?) *Coronation of the Virgin*, Rome, S. Onofrio

93. PERUZZI *Decoration of the Chapel*, Castello di Belcaro

94. PERUZZI *Madonna with Saints* (altar fresco), Rome, S. Onofrio

95. RIPANDA *Consul and Lictors*, Rome, Palazzo dei Conservatori

96. SODOMA *Central Octagon of the Stanza della Segnatura*, Rome, Vatican

III
THE MATURITY OF THE CLASSICAL STYLE IN ROME
(c. 1508-c. 1514)

97. MICHELANGELO *The Sistine Ceiling*, Rome, Vatican, Sistine Chapel

98. MICHELANGELO *The Sistine Ceiling* (central bays)

99. MICHELANGELO *Preliminary Plan for Sistine Ceiling*, London, British Museum

100. MICHELANGELO *Preliminary Plan for Sistine Ceiling*, Detroit, Institute of Arts

101, 102. MICHELANGELO *Bronze-colored Nudes in the Spandrels*, Sistine Ceiling

103. MICHELANGELO *The Flood*, Sistine Ceiling

104. MICHELANGELO *Sacrifice of Noah*, Sistine Ceiling

105. MICHELANGELO *Drunkenness of Noah*, Sistine Ceiling

107. MICHELANGELO *Prophet Joel*, Sistine Ceiling

106. MICHELANGELO *Delphic Sibyl*, Sistine Ceiling

109. MICHELANGELO *Erithrean Sibyl*, Sistine Ceiling

108. MICHELANGELO *Prophet Zachary*, Sistine Ceiling

110. MICHELANGELO *Prophet Isaiah*, Sistine Ceiling

111. MICHELANGELO *Ignudi around the Drunkenness of Noah*, Sistine Ceiling

112, 113. MICHELANGELO *Ignudi above Prophet Joel*, Sistine Ceiling

114. MICHELANGELO *Ignudi around Sacrifice of Noah*, Sistine Ceiling

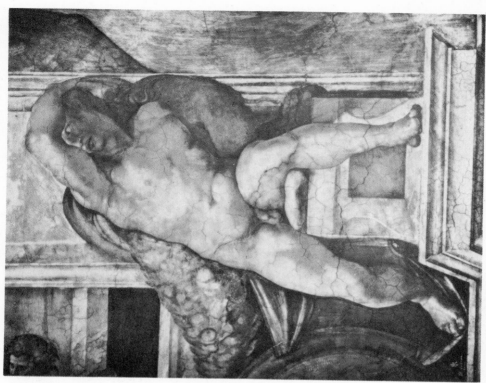

115, 116. MICHELANGELO *Ignudi above Prophet Isaiah*, Sistine Ceiling

117. MICHELANGELO *Creation of Eve*, Sistine Ceiling

118. MICHELANGELO *Ignudi around Creation of Eve*, Sistine Ceiling

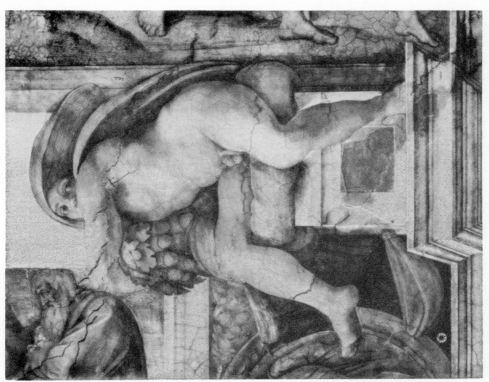
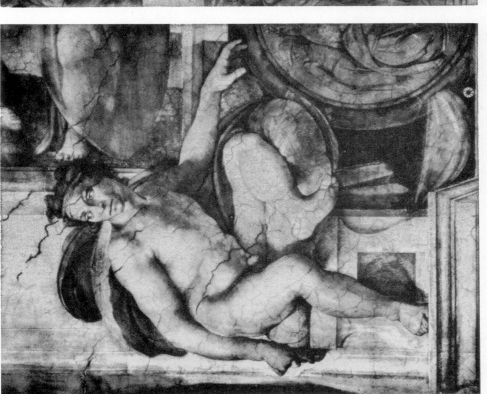

119, 120. MICHELANGELO *Ignudi above Cumaean Sibyl*, Sistine Ceiling

121. MICHELANGELO *Temptation and Expulsion*, Sistine Ceiling

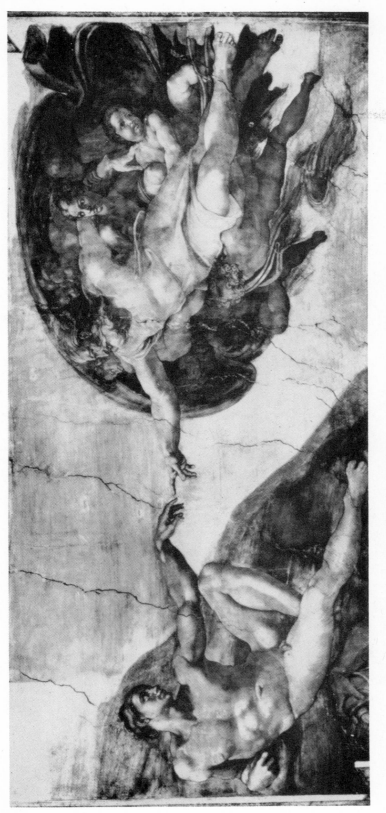

122. MICHELANGELO *Creation of Adam*, Sistine Ceiling

123. MICHELANGELO *Creation of Adam* (detail), Sistine Ceiling

124. MICHELANGELO *Creation of the Sun and Moon*, Sistine Ceiling

125. MICHELANGELO *Separation of Earth and Waters*, Sistine Ceiling

126. MICHELANGELO *Separation of Light and Darkness*, Sistine Ceiling

127. **MICHELANGELO** *Ignudi around Separation of Earth and Waters*, Sistine Ceiling

128, 129. MICHELANGELO *Ignudi above Persian Sibyl*, Sistine Ceiling

130. **MICHELANGELO** *Ignudi around Separation of Light and Darkness*, Sistine Ceiling

131, 132. MICHELANGELO *Ignudi above Prophet Jeremiah, Sistine Ceiling*

133. MICHELANGELO *Cumaean Sibyl*, Sistine Ceiling

134. MICHELANGELO *Persian Sibyl*, Sistine Ceiling

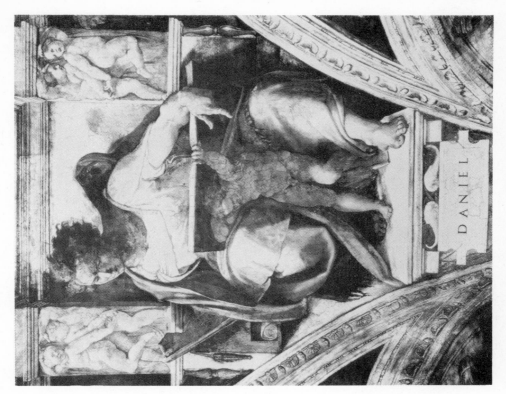

136. MICHELANGELO *Prophet Daniel*, Sistine Ceiling

135. MICHELANGELO *Prophet Ezekiel*, Sistine Ceiling

137. MICHELANGELO *Libyan Sibyl*, Sistine Ceiling

138. MICHELANGELO *Prophet Jonah*, Sistine Ceiling

139. **MICHELANGELO** *Prophet Jeremiah*, Sistine Ceiling

140. MICHELANGELO *David and Goliath*, Sistine Ceiling

141. MICHELANGELO *Judith and Holofernes*, Sistine Ceiling

142. MICHELANGELO *The Hanging of Haman*, Sistine Ceiling

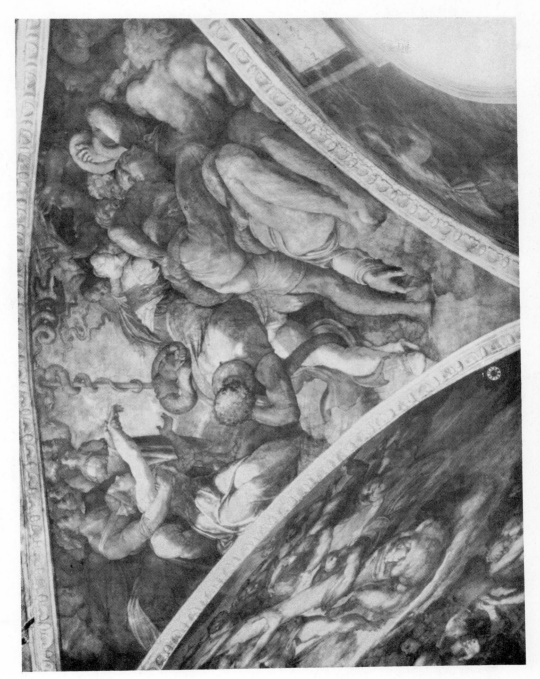

143. MICHELANGELO *The Brazen Serpent*, Sistine Ceiling

144. MICHELANGELO *Ancestors of Christ*, Sistine Ceiling, severies

146. MICHELANGELO *Ancestors of Christ*, Sistine Ceiling, lunettes

145. MICHELANGELO *Ancestors of Christ*, Sistine Ceiling, severies

147. MICHELANGELO *Ancestors of Christ*, Sistine Ceiling, lunettes

148. MICHELANGELO *Ancestors of Christ*, Sistine Ceiling

149. RAPHAEL *Stanza della Segnatura*, Rome, Vatican

150. RAPHAEL *Stanza della Segnatura*

151. RAPHAEL [and Sodoma] *Ceiling of the Stanza della Segnatura*

152. RAPHAEL *Poetry*, Stanza della Segnatura, ceiling

153. RAPHAEL *Justice*, Stanza della Segnatura, ceiling

154. RAPHAEL *The Flaying of Marsyas*, Stanza della Segnatura, ceiling

155. RAPHAEL *The Judgment of Solomon*, Stanza della Segnatura, ceiling

156. RAPHAEL *Disputà*, Stanza della Segnatura

157. RAPHAEL *Disputà* (detail)

158. RAPHAEL *Disputà* (detail)

159. RAPHAEL *Parnassus*, Stanza della Segnatura

160. RAPHAEL *Parnassus* (detail), *Apollo and Muses*

161. RAPHAEL *Parnassus* (detail), *Sappho and Other Poets*

162. RAPHAEL *Parnassus* (detail), *Modern Poets*

163. RAPHAEL *School of Athens*, Stanza della Segnatura

164. RAPHAEL *School of Athens* (detail), *Plato and Aristotle*

165. RAPHAEL *School of Athens* (detail), *Pythagorean Group*

166. RAPHAEL *School of Athens* (detail), *Euclidian Group*

167. RAPHAEL
School of Athens (detail), *Heraclitus*

168. RAPHAEL *Cartoon for School of Athens* (detail), Milan, Ambrosiana

169. RAPHAEL *Cartoon for School of Athens*, Milan, Ambrosiana

170. RAPHAEL *The Law*, Stanza della Segnatura

171. RAPHAEL *The Three Virtues of Justice*, Stanza della Segnatura

172. RAPHAEL *Handing over of the Decretals*, Stanza della Segnatura

173. RAPHAEL *The Civil Law*, Stanza della Segnatura

174. RAPHAEL [with Penni] *Grisaille and Grotesque Decoration* (beneath *Parnassus*), Stanza della Segnatura

175. RAPHAEL   *Madonna di Casa Alba*, Washington, National Gallery

176. RAPHAEL   *Study for Alba Madonna*, Lille, Musée Wicar

177. RAPHAEL *Madonna di Foligno*, Rome, Vatican Museum

178. RAPHAEL *Isaiah*, Rome, S. Agostino

179. RIPANDA (?) *Triumph of Titus*, Paris, Louvre
PERUZZI (?)

180. [Anon.] *Façade Decoration of the Casa Sander*, Rome

181. [Anon.] *Façade Decoration*, Via Maschera d'Oro no. 9, Rome

182. PERUZZI *Three Graces*, San Francisco, Zellerbach Collection

183. PERUZZI *Sala di Galatea*, Rome, Villa Farnesina

184. PERUZZI *Aquarius between the Swan and Dolphin.* SEBASTIANO *Tireus and Philomel; The Daughters of Cecrops,* Sala di Galatea

185. PERUZZI *Perseus-Pegasus*, Sala di Galatea

186. PERUZZI *Ursa Major*, Sala di Galatea

187. PERUZZI *Luna in Virgo, Bacchus and Ariadne, Mars in Libra near Scorpio.*
     SEBASTIANO *Fall of Phaeton,* Sala di Galatea

188. PERUZZI *Venus in Capricorn with Sagittarius and Lyra,*
     Sala di Galatea

189. PERUZZI *Argo*, Sala di Galatea

190. PERUZZI *Sol in Sagittarius*, Sala di Galatea

191. PERUZZI *Death of Meleager*, Rome, Villa Farnesina, Sala del Fregio

192. PERUZZI *Hunting of Calydonian Boar*, Sala del Fregio

193. PERUZZI *Nymph and Satyrs; Slaying of Marsyas* (portions), Sala del Fregio

194. SODOMA *Marriage of Alexander and Roxane*, Rome, Villa Farnesina

195. SODOMA *Alexander and the Family of Darius; The Forge of Vulcan*, Rome, Villa Farnesina

196. SEBASTIANO *Juno*, Sala di Galatea, Rome, Villa Farnesina

197. SEBASTIANO *Fall of Icarus*, Sala di Galatea

198. SEBASTIANO *Polyphemus*, Sala di Galatea

199. SEBASTIANO *Adoration of the Shepherds*, Cambridge, Fitzwilliam Museum

200. SEBASTIANO *Death of Adonis*, Florence, Uffizi

201. SEBASTIANO *Madonna and Child*, London, Pouncey Collection

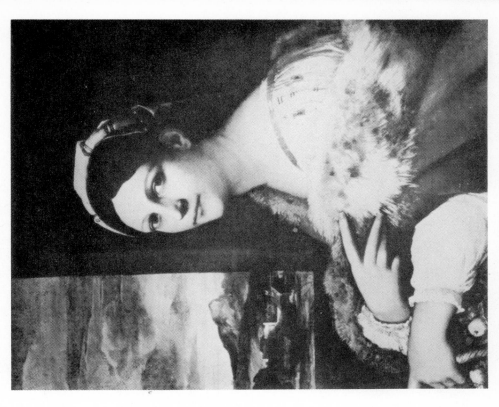

203. SEBASTIANO *Portrait of a Girl* (called "Dorothea"), Berlin, Museums

202. SEBASTIANO *Portrait of a Girl* (called "La Fornarina"), Florence, Uffizi

204. PERUZZI *Giant Head*, Sala di Galatea, Rome, Villa Farnesina

205. PERUZZI *Ceiling of the Stanza d'Eliodoro*, Rome, Vatican
RAPHAEL and PERUZZI

206. PERUZZI (?)   *Project for Stanza d'Eliodoro*, Paris, Louvre
     PENNI (?) after RAPHAEL

207. PERUZZI   *Study for Ceiling*, Stanza d'Eliodoro; Oxford, Ashmolean
     PERUZZI after RAPHAEL

208. PERUZZI *Cartoon for Moses and the Burning Bush*, Stanza d'Eliodoro; Naples, Galleria Nazionale
RAPHAEL (?)

209. PERUZZI *Moses and the Burning Bush*, Stanza d'Eliodoro, ceiling
RAPHAEL (assisted?)

210. PERUZZI *Jacob's Dream*, Stanza d'Eliodoro, ceiling RAPHAEL (assisted?)

212. PERUZZI *Holy Family in a Landscape*, London, Pouncey Collection

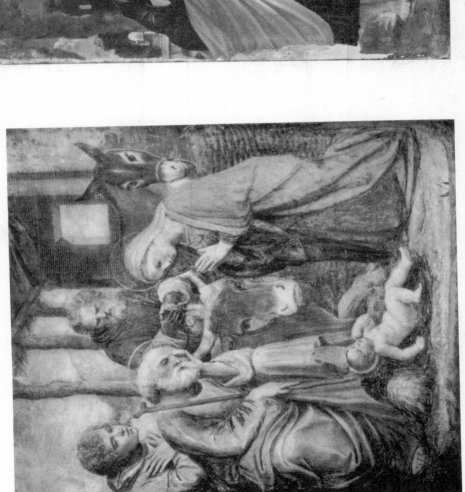

211. PERUZZI *Adoration of the Child*, Rome, S. Rocco

213. RAPHAEL Stanza d'Eliodoro, Rome, Vatican

214. RAPHAEL *Mass of Bolsena*, Stanza d'Eliodoro

215. RAPHAEL *Mass of Bolsena* (detail)

216. RAPHAEL *Mass of Bolsena* (detail), *Swiss Guards*

217. RAPHAEL *Expulsion of Heliodorus*, Stanza d'Eliodoro

218. RAPHAEL *Expulsion of Heliodorus* (detail)

220. RAPHAEL Repulse of Attila (detail)

219. RAPHAEL Expulsion of Heliodorus (detail)

221. RAPHAEL Repulse of Attila, Stanza d'Eliodoro

222. RAPHAEL *Liberation of Peter*, Stanza d'Eliodoro

223. RAPHAEL *Liberation of Peter* (detail)

224. RAPHAEL *Liberation of Peter* (detail)

226. GIULIO *Basamento* (detail; repainted), Stanza d'Eliodoro

225. RAPHAEL ASSISTANT *Grotesque Decoration*, Stanza d'Eliodoro

227. RAPHAEL *Study for a Resurrection*, Bayonne, Musée Bonnat

228. RAPHAEL   *Study for a Resurrection*, Windsor, Royal Library

229. RAPHAEL   *Study for a Resurrection*, Oxford, Ashmolean

230. RAPHAEL *Galatea*, Rome, Villa Farnesina

231. RAPHAEL ASSISTANT *Prophets*, Rome, S.M. della Pace, Cappella Chigi

232. RAPHAEL *Sibyls*, S.M. della Pace, Cappella Chigi

233. RAPHAEL *Sibyls* (detail), S.M. della Pace

234. RAPHAEL *Sibyls* (detail), S.M. della Pace

235. RAPHAEL *Sistine Madonna*, Dresden, Gallery

237. RAPHAEL *Sistine Madonna* (detail), *St. Barbara*

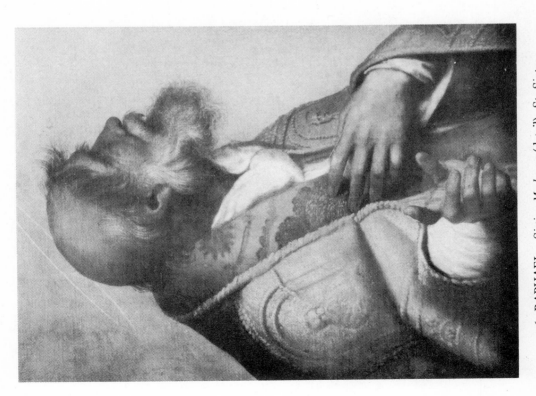

236. RAPHAEL *Sistine Madonna* (detail), *St. Sixtus*

239. RAPHAEL *Study for Madonna dell'Impannata*, Windsor, Royal Library

238. RAPHAEL [with Penni and Giulio] *Madonna of the Fish*, Madrid, Prado

240. RAPHAEL [with Giulio]   *Madonna dell'Impannata*, Florence, Pitti

241. RAPHAEL *St. Cecilia Altar*, Bologna, Pinacoteca

243. RAPHAEL *St. Cecilia Altar* (detail)

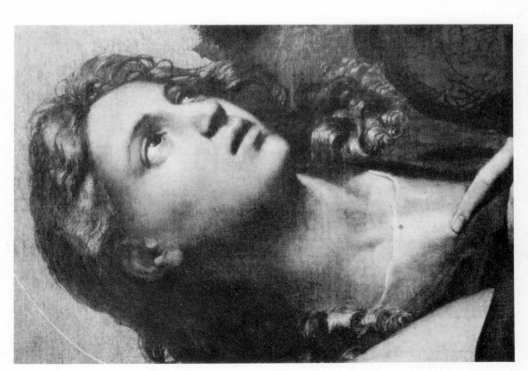

242. RAPHAEL *St. Cecilia Altar* (detail)

245. RAPHAEL ASSISTANT *Tommaso Inghirami*, Florence, Pitti

244. RAPHAEL *Tommaso Inghirami* ("Il Fedra"), Boston, Gardner Museum

247. RAPHAEL. *La Donna Velata*, Florence, Pitti

246. RAPHAEL [or assistant] *Giuliano de' Medici*, New York, Metropolitan Museum

248. RAPHAEL *Madonna della Sedia*, Florence, Pitti

249. RAPHAEL *Madonna della Sedia* (detail)

250. LEONARDO [and assistant ?]
St. John Baptist, Paris, Louvre

251. LEONARDO *Cataclysm*, Windsor, Royal Library

IV
THE MATURATION OF CLASSICAL STYLE IN FLORENCE (c. 1508-c. 1514)

252. FRA BARTOLOMMEO *Holy Family*, London, National Gallery 3914

253. FRA BARTOLOMMEO *God the Father with Sts. Mary Magdalen and Catherine of Siena*, Lucca, Pinacoteca

254. FRA BARTOLOMMEO [with Albertinelli] *Madonna with Six Saints*, Florence, S. Marco

255. FRA BARTOLOMMEO *Madonna with Sts. Stephen and John Baptist*, Lucca, Cathedral

256. ALBERTINELLI *Madonna with Four Saints*, Florence, Academy

258. ALBERTINELLI *Study for a Trinity*, Florence, Uffizi

257. ALBERTINELLI *Trinity*, Florence, Academy

259. ALBERTINELLI *Annunciation*, Florence, Academy

260. FRA BARTOLOMMEO *Marriage of St. Catherine*, Paris, Louvre

261. FRA BARTOLOMMEO [with Albertinelli] *Virgin in Glory with Saints*, Besançon, Cathedral

262. ALBERTINELLI *Coronation of the Virgin* (fragment; former crown-piece of 261), Stuttgart, Gallery

263. FRA BARTOLOMMEO *St. Anne Altar*, Florence, Museo di S. Marco

264. FRA BARTOLOMMEO [with assistants] *The Marriage of St. Catherine* (The *Pitti Pala*), Florence, Academy

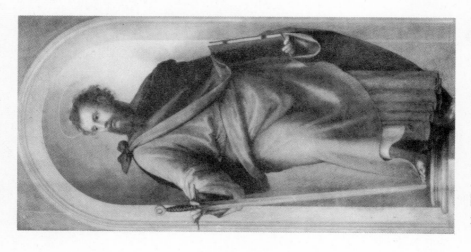

266. FRA BARTOLOMMEO *St. Paul*, Rome, Vatican Museum

265. FRA BARTOLOMMEO [with Raphael] *St. Peter*, Rome, Vatican Museum

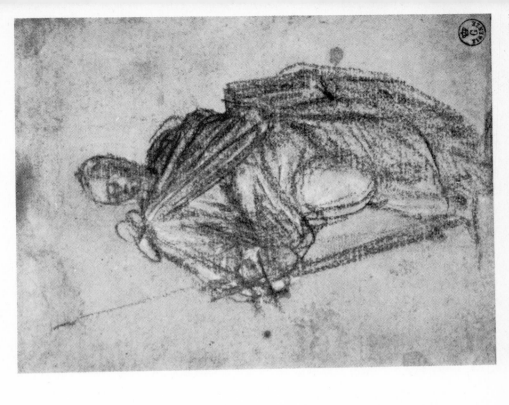

268. FRA BARTOLOMMEO *Study for St. Paul*, Florence, Uffizi

267. FRA BARTOLOMMEO [with Raphael] *St. Peter* (detail)

270. BUGIARDINI *Portrait of a Young Woman*, Paris, Musée Jacquemart-André

269. BUGIARDINI *La Monaca*, Florence, Pitti

272. BUGIARDINI *Madonna and Child with St. John* (formerly ?) New York, C. H. Holmes Collection

271. BUGIARDINI *Madonna Standing in a Landscape* (sold London, 1946)

273. BUGIARDINI *Madonna and Child with St. John* (formerly) London, Agnew

274. BUGIARDINI *Madonna del Latte*, Florence, Uffizi

275. BUGIARDINI   *Ariadne* (?), Venice, Ca' d'Oro

276. BUGIARDINI   *Leda*, Milan, Treccani Collection

277. RIDOLFO GHIRLANDAIO
Portrait of a Lady, Florence, Pitti

278. RIDOLFO GHIRLANDAIO
Adoration of the Child
(destroyed; formerly Berlin, Museums)

279. RIDOLFO GHIRLANDAIO *Adoration of the Shepherds*, Budapest, Museum

280. RIDOLFO GHIRLANDAIO *Nativity with Six Saints*, New York, Metropolitan Museum

281. RIDOLFO GHIRLANDAIO [shop assistant]
Adoration of the Shepherds (formerly) London, Henry Harris Collection

282. RIDOLFO GHIRLANDAIO *Madonna della Cintola*, Prato, Cathedral

283. RIDOLFO GHIRLANDAIO *Portrait of a Goldsmith*, Florence, Pitti

284. RIDOLFO GHIRLANDAIO [with Andrea di Cosimo] *Decoration of Cappella dei Priori*, Florence, Palazzo Vecchio

285. RIDOLFO GHIRLANDAIO [with Andrea di Cosimo] *Decoration of Cappella dei Priori*

286. GRANACCI *Madonna with Two Saints*, Villamagna (near Florence), S. Donnino

287. GRANACCI *Madonna della Cintola*, Sarasota, Ringling Museum

289. GRANACCI *Pietà*, Quintole, S. Pietro

288. GRANACCI *Trinity*, Berlin, Museums

290. GRANACCI *Madonna in Glory with Four Saints*, Florence, Academy

291. GRANACCI *Holy Family with St. John*, Florence, Pitti

292. PIERO DI COSIMO [with assistants] *Doctrine of the Immaculate Conception*, Fiesole, S. Francesco

293. PIERO DI COSIMO   *Legend of Prometheus*, Munich, Pinakothek

294. PIERO DI COSIMO   *Legend of Prometheus*, Strasbourg, Museum

295. PIERO DI COSIMO   *Adoration of the Child*, Rome, Borghese

296. ANDREA DEL SARTO *Pietà*, Rome, Borghese

297. ANDREA DEL SARTO *Madonna*, Rome, Galleria Nazionale

298. ANDREA DEL SARTO *Baptism of Christ*, Florence, Scalzo

300. ANDREA DEL SARTO *Burial of St. Philip*, Florence, SS. Annunziata

299. ANDREA DEL SARTO *Healing of the Obsessed Girl*, Florence, SS. Annunziata

301. ANDREA DEL SARTO *Healing by St. Philip's Relics*, Florence, SS. Annunziata

302. ANDREA DEL SARTO *Punishment of the Gamblers*, Florence, SS. Annunziata

303. ANDREA DEL SARTO *Clothing of the Leper*, Florence, SS. Annunziata

304. ANDREA DEL SARTO *Noli Me Tangere*, Florence, Uffizi

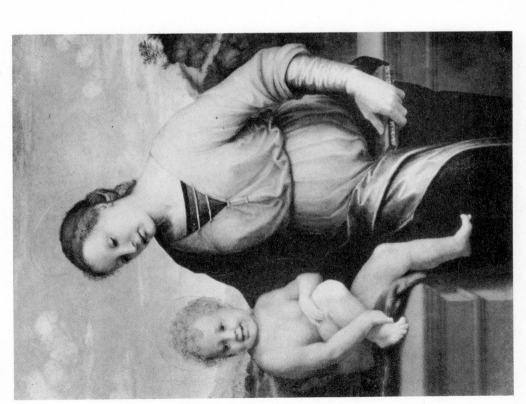

305. FRANCIABIGIO *Madonna*, Rome, Galleria Nazionale

306. FRANCIABIGIO *Adoration of the Shepherds*, Florence, Museo di S. Marco

307. FRANCIABIGIO *Last Supper*, Florence, S.M. dei Candeli

308. ANDREA DEL SARTO *Adoration of the Magi*, Florence, SS. Annunziata

309. ANDREA DEL SARTO *Annunciation*, Florence, Pitti

310. ANDREA DEL SARTO [with Puligo ?]   *Marriage of St. Catherine*, Dresden, Gallery
ANDREA DEL SARTO

311. ANDREA DEL SARTO *Birth of the Virgin*, Florence, SS. Annunziata

312. ANDREA DEL SARTO *Birth of the Virgin* (detail)

313. ANDREA DEL SARTO *Birth of the Virgin* (detail)

314. FRANCIABIGIO *Marriage of the Virgin*, Florence, SS. Annunziata

315. FRANCIABIGIO *Last Supper*, Florence, Convento della Calza

316. FRANCIABIGIO *Last Supper* (detail)

318. FRANCIABIGIO *Portrait of a Man*, London, National Gallery

317. FRANCIABIGIO *Portrait of a Man*, Florence, Uffizi 8381

320. ANDREA DEL SARTO [with Puligo ?] *Madonna with the Infant St. John*, Rome, Borghese 336

319. PULIGO *Madonna and Child with St. John*, Rome, Palazzo Venezia

321. ANDREA DEL SARTO [with Puligo?] *Tobias Altar*, Vienna, Kunsthistorisches Museum
ANDREA DEL SARTO

322. PULIGO *Madonna with St. John Approaching in a Landscape*, Rome, Borghese 338

323. PONTORMO *Ospedale di S. Matteo*, Florence, Academy

324. PONTORMO *Madonna with Four Saints* (altar fresco from S. Ruffillo), Florence, SS. Annunziata

325. ROSSO *Madonna and Child* New York, Finch College, Kress Collection 485
MASTER OF THE KRESS LANDSCAPES

326. ROSSO *Madonna in a Landscape* Arezzo, Museum (from Uffizi Deposit 8309)
MASTER OF THE KRESS LANDSCAPES

328. BERRUGUETE *Madonna and Elizabeth with the Two Holy Children*, Rome, Borghese 335

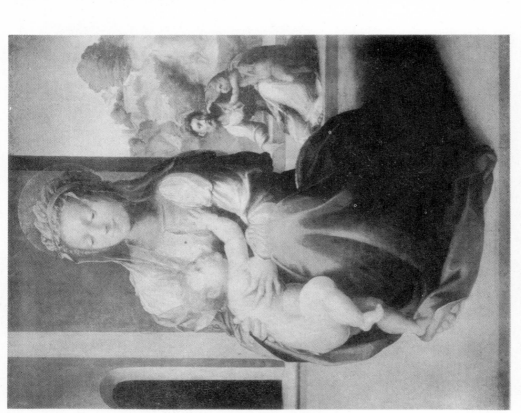

327. ROSSO *Holy Family*, Rome, Borghese

MASTER OF THE KRESS LANDSCAPES

330. [Anon.] *Madonna* (formerly) Milan, Crespi Collection

329. BERRUGUETE *Madonna*, Milan, Saibene Collection

331. FILIPPINO LIPPI, BERRUGUETE [and others]   *Coronation of the Virgin*, Paris, Louvre

332. MANCHESTER MASTER *Madonna and Child with St. John*, Vienna, Academy

333. MANCHESTER MASTER *Virgin Reading with the Christ Child and St. John*, New York, Kress Collection

335. MANCHESTER MASTER *Madonna*, Baden bei Zurich, Private Collection

334. MANCHESTER MASTER *Pietà*, Rome, Galleria Nazionale

337. MANCHESTER MASTER *Madonna*, Florence, Art Market

336. MANCHESTER MASTER *Madonna with St. John and Four Angels*, London, National Gallery

338. MICHELANGELO [with the Manchester Master]   *Entombment*, London National Gallery

V
CLIMAX, CRISIS, AND DISSOLUTION OF THE CLASSICAL STYLE IN ROME
(c. 1514-c. 1520)

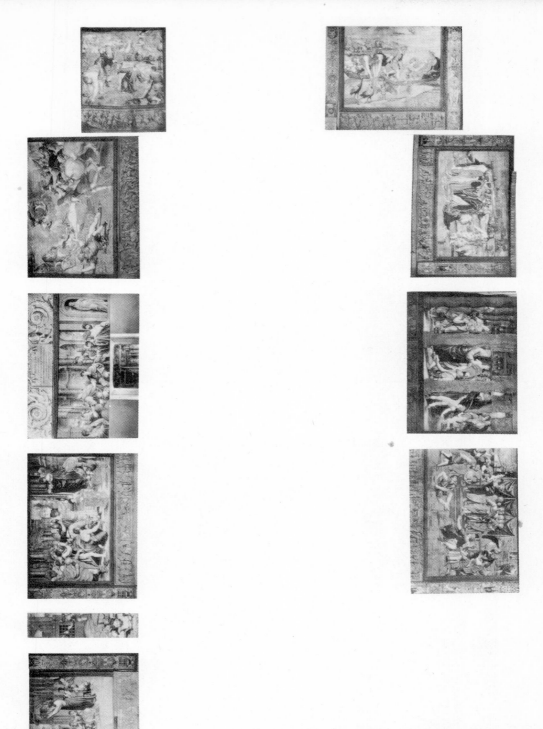

339. RAPHAEL *Acts of the Apostles* (tapestries), Rome, Vatican Museum (in order of arrangement in the Sistine Chapel)

340. RAPHAEL *The Stoning of Stephen* (tapestry)

341. RAPHAEL *The Conversion of Paul* (tapestry)

342. RAPHAEL *Miraculous Draught of Fishes* (tapestry)

343. RAPHAEL *Pasce Oves* (tapestry)

344. RAPHAEL *Blinding of Elymas* (tapestry fragment)

345. RAPHAEL *Paul at Lystra* (tapestry)

346. RAPHAEL *Healing at the Golden Gate* (tapestry)

347. RAPHAEL *Death of Ananias* (tapestry)

348. RAPHAEL *Paul Preaching at Athens* (tapestry)

349. RAPHAEL *Study for the Pasce Oves*, Windsor, Royal Library

350. RAPHAEL *Study for the Christ of the Pasce Oves*, Paris, Louvre

351. RAPHAEL *Study for the Paul at Lystra*, Chatsworth

352. RAPHAEL *Study for the Blinding of Elymas*, Windsor, Royal Library

353. PENNI *Study for the Pasce Oves*, Paris, Louvre
RAPHAEL (?)

354. PENNI *Study for Paul Preaching at Athens*, Florence, Uffizi

355. RAPHAEL *Miraculous Draught of Fishes* (tapestry cartoon), London, Victoria and Albert Museum

356. RAPHAEL *Miraculous Draught of Fishes* (tapestry cartoon, detail)

357. RAPHAEL *Miraculous Draught of Fishes* (tapestry cartoon, detail)

358. RAPHAEL *Pasce Oves* (tapestry cartoon), London, Victoria and Albert Museum

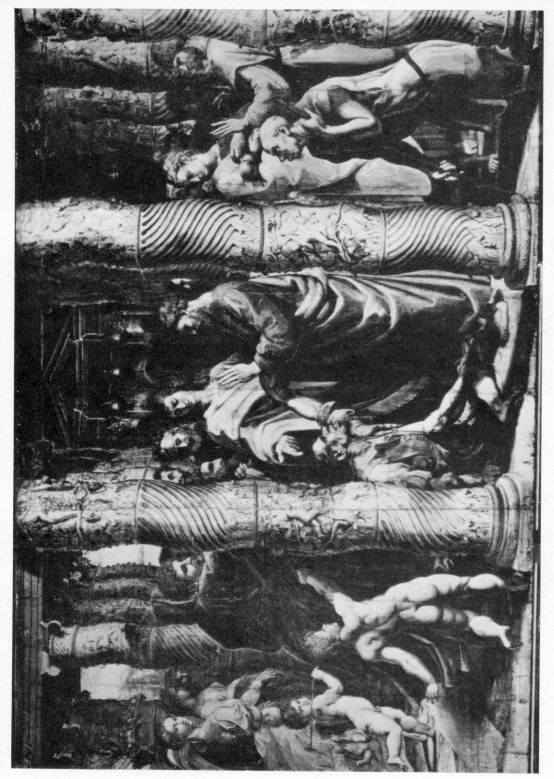

359. RAPHAEL *Healing at the Golden Gate* (tapestry cartoon), London, Victoria and Albert Museum

361. RAPHAEL *Healing at the Golden Gate* (tapestry cartoon, detail)

360. RAPHAEL *Healing at the Golden Gate* (tapestry cartoon, detail)

362. RAPHAEL *Death of Ananias* (tapestry cartoon), London, Victoria and Albert Museum

363. RAPHAEL *Death of Ananias* (tapestry cartoon, detail)

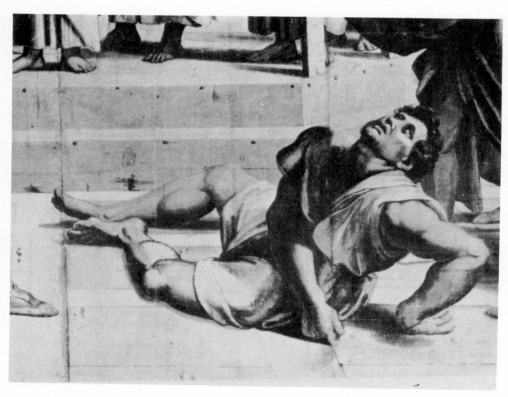

364, 365. RAPHAEL *Death of Ananias* (tapestry cartoon, details)

366. RAPHAEL [with Penni] *Blinding of Elymus* (tapestry cartoon), London, Victoria and Albert Museum

367. RAPHAEL *Paul at Lystra* (tapestry cartoon), London, Victoria and Albert Museum

368. RAPHAEL *Paul at Lystra* (tapestry cartoon, detail)

369. RAPHAEL [with Penni] *Paul Preaching at Athens* (tapestry cartoon), London, Victoria and Albert Museum

370. PENNI *Paul Preaching at Athens* (tapestry cartoon, detail)

371. RAPHAEL *Paul at Lystra* (tapestry cartoon, detail)

372. PENNI *Study for the Miraculous Draught of Fishes*, Vienna, Albertina
    GIULIO

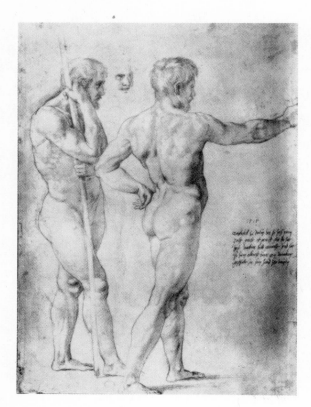

373. GIULIO [retouched by Raphael?]
    Study for the Battle of Ostia, Vienna, Albertina
    RAPHAEL

374. RAPHAEL [with Giulio] *Battle of Ostia*, Rome, Vatican, Stanza dell'Incendio

375. GIULIO *Battle of Ostia* (detail)

376. RAPHAEL [with Giulio] *Fire in the Borgo*, Rome, Vatican, Stanza dell'Incendio

377. GIULIO *Fire in the Borgo* (detail)

378, 379. GIULIO *Fire in the Borgo* (details)

380. RAPHAEL [with Penni] *Coronation of Charlemagne*, Rome, Vatican, Stanza dell'Incendio

381. PENNI *Coronation of Charlemagne* (detail)

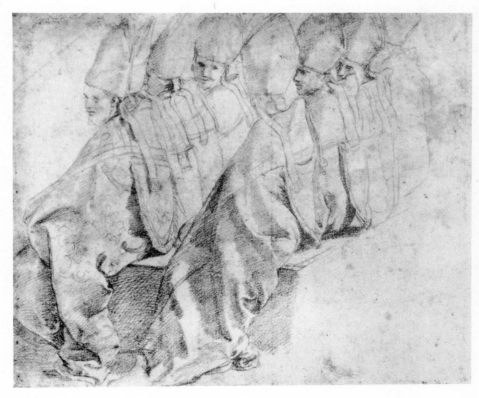

382. PENNI *Study for the Coronation of Charlemagne*, Düsseldorf, Museum

383. RAPHAEL [with Penni] *Oath of Leo*, Rome, Vatican, Stanza dell'Incendio

384. PENNI *Study for the Oath of Leo*, Florence, Horne Foundation

385. GIULIO *Basamento* (detail), Rome, Vatican, Stanza dell'Incendio

386. RAPHAEL [with Alvise de Pace] *Cupola of the Cappella Chigi*, Rome, S.M. del Popolo

387, 388. RAPHAEL *Studies for the Cappella Chigi*, Oxford, Ashmolean

389. RAPHAEL [with Giovanni da Udine] *Loggetta of the Cardinal Bibbiena*, Rome, Vatican

390. RAPHAEL [with Giovanni da Udine] *Loggetta of the Cardinal Bibbiena*, Rome, Vatican

391. GIOVANNI DA UDINE [and assistants]   *Loggetta of the Cardinal Bibbiena* (detail)

392, 393. RAPHAEL [with Giovanni da Udine] *Stufetta of the Cardinal Bibbiena*, Rome, Vatican

394. TADDEO ZUCCARO [and others] *Decoration of the Sala dei Palafrenieri* (free reconstruction of Raphael), Rome, Vatican

395, 396. PENNI *Studies for the Sala dei Palafrenieri*, Paris, Louvre

397. RAPHAEL [and assistants] *Sala di Psiche*, Rome, Villa Farnesina

398. GIULIO *Three Graces*, Sala di Psiche

399. GIULIO *Venus, Ceres, and Juno*, Sala di Psiche

400. GIULIO *Jupiter and Cupid*, Sala di Psiche

401. GIULIO *Venus and Psyche*, Sala di Psiche

402. RAPHAEL *Study for Venus and Psyche*, Paris, Louvre

404. PENNI *Mercury Descending*, Sala di Psiche

403. PENNI *Venus before Jupiter*, Rome, Villa Farnesina, Sala di Psiche

406. GIOVANNI DA UDINE *Amoretto with Mythic Beasts*, Sala di Psiche

405. RAFFAELLINO DEL COLLE (?) *Venus and Cupid*, Sala di Psiche

407. RAPHAEL Study for the Wedding Feast of Cupid and Psyche, Windsor, Royal Library

408. PENNI *Council of the Gods*, Rome, Villa Farnesina, Sala di Psiche

409. PENNI *Wedding Feast of Cupid and Psyche*, Sala di Psiche

410. RAPHAEL [and assistants]   *Logge*, Rome, Vatican

411, 412. GIOVANNI DA UDINE [and assistants] *Grotesque Decorations* (details), Rome, Vatican, Logge

413. GIOVANNI DA UDINE *Borders from the Tapestries of the Acts of the Apostles*, Rome, Vatican Museum

414. POLIDORO *Grotesque Decorations*, Rome, Vatican, Stanza dell'Incendio

415. RAPHAEL [and assistants]   *Logge* (first bay), Rome, Vatican

416. RAPHAEL [and assistants]   *Logge* (second bay), Rome, Vatican

417. RAPHAEL [and assistants]   *Logge* (third bay), Rome, Vatican

418. RAPHAEL [and assistants]   *Logge* (fourth bay), Rome, Vatican

419. RAPHAEL [and assistants] *Logge* (fifth bay), Rome, Vatican

420. RAPHAEL [and assistants] *Logge* (sixth bay), Rome, Vatican

421. RAPHAEL [and assistants]   *Logge* (seventh bay), Rome, Vatican

422. GIOVANNI DA UDINE   *Lower Loggia*, Rome, Vatican, Cortile di S. Damaso

423. RAPHAEL   *Baldassare Castiglione*, Paris, Louvre

425. RAPHAEL *Antonio Tebaldeo* (copy), Florence, Uffizi

424. RAPHAEL *Antonio Tebaldeo* (from the *Parnassus* fresco), Rome, Vatican

426. RAPHAEL *Andrea Navagero and Agostino Beazzano*, Rome, Galleria Doria

428. RAPHAEL *Cardinal Bernardo Bibbiena* (copy), Florence, Uffizi

427. GIULIO *Bindo Altoviti*, Washington, National Gallery, Kress Collection

429. RAPHAEL   *Leo X with the Cardinals Giulio de' Medici and Luigi Rossi*, Florence, Pitti

430. RAPHAEL *Raphael and His Fencing Master*, Paris, Louvre

431. GIULIO *Giovanna d'Aragona*, Paris, Louvre

432. RAPHAEL *Madonna della Tenda*, Munich, Pinakothek

433. RAPHAEL [with Penni]   *Spasimo di Sicilia*, Madrid, Prado

435. GIULIO *Cartoon for the Holy Family of Francis I* (fragment), Melbourne, National Gallery

434. RAPHAEL *Spasimo di Sicilia* (detail)

436. RAPHAEL [with Giulio] *Holy Family of Francis I*, Paris, Louvre

437. RAPHAEL [with Giulio] *St. Michael*, Paris, Louvre

438. RAPHAEL [and assistants] *Transfiguration*, Rome, Vatican Museum

439. PENNI *Transfiguration* (detail)

440. RAPHAEL [and Giulio] *Transfiguration* (detail)

441. GIULIO *Transfiguration* (detail)

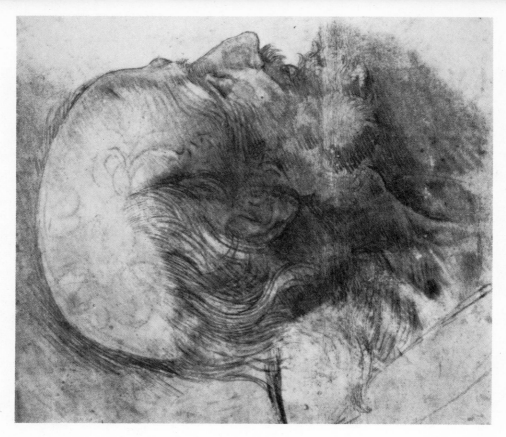

443. RAPHAEL *Study for St. Andrew*, London, British Museum

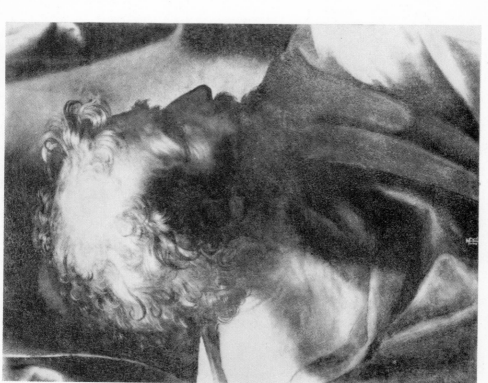

442. RAPHAEL *Transfiguration* (detail), St. Andrew

445. GIULIO (?) *Study for the Transfiguration,* Paris, Louvre

444. RAPHAEL *Study for the Transfiguration,* Oxford, Ashmolean

446. GIULIO *Madonna Piccola Gonzaga*, Paris, Louvre

447. GIULIO *St. Margaret*, Paris, Louvre

448. GIULIO *Madonna della Perla*, Madrid, Prado

450. PENNI *Madonna del Divino Amore*, Naples, Galleria Nazionale

449. PENNI *Visitation*, Madrid, Prado

452. GIULIO [with Raffaellino] *Madonna della Rosa*, Madrid, Prado

451. GIULIO [with Penni] *St. John Baptist*, Florence, Academy

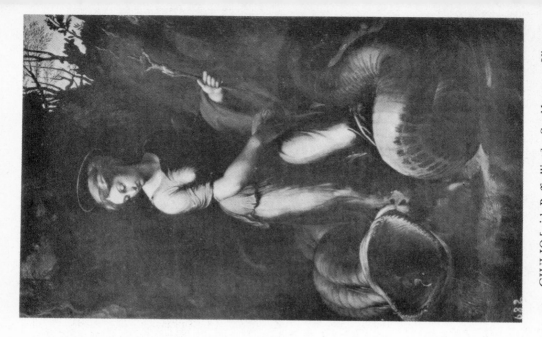

453. GIULIO [with Raffaellino] *Madonna of the Oak*, Madrid, Prado

454. GIULIO [with Raffaellino] *St. Margaret*, Vienna, Kunsthistorisches Museum

455. SEBASTIANO *Man in Armor*, Hartford, Atheneum

457. SEBASTIANO *Portrait of a Young Man*, Budapest, Museum

456. SEBASTIANO *Young Violinist*, Paris, Baron G. de Rothschild

458. SEBASTIANO *Cardinal Antonio Ciocchi del Monte Sansovino*, Dublin, National Gallery

459. SEBASTIANO *Verdelotti and Ubretto* (destroyed; formerly Berlin, Museums)

460. SEBASTIANO *Cardinal Bandinello Sauli and Suite*, Washington, National Gallery, Kress Collection

461. SEBASTIANO *Pietà*, Viterbo, Museum

462. SEBASTIANO *Pietà*, Leningrad, Hermitage

463. SEBASTIANO *Resurrection of Lazarus*, London, National Gallery

464. SEBASTIANO *Resurrection of Lazarus* (detail)

466. SEBASTIANO *Holy Family with St. John Baptist and Donor*, London, National Gallery

465. SEBASTIANO *Resurrection of Lazarus* (detail)

467. SEBASTIANO *Cappella Borgherini*, Rome, S. Pietro in Montorio

468. SEBASTIANO *Two Prophets*, Cappella Borgherini

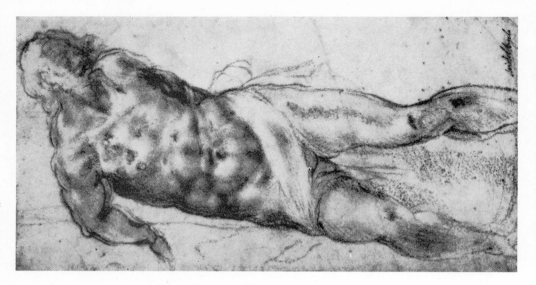

470. SEBASTIANO *Study for the Flagellation*, London, British Museum

469. SEBASTIANO *Flagellation*, Cappella Borgherini

471. MICHELANGELO *Modello for the Flagellation* (copy), Windsor, Royal Library

472. SEBASTIANO *Study for the Flagellation*, London, British Museum

473. SEBASTIANO *Transfiguration*, Cappella Borgherini

474. SEBASTIANO *Visitation*, Paris, Louvre

475. SEBASTIANO *Martyrdom of St. Agatha*, Florence, Pitti

476. SEBASTIANO *Study for St. Agatha*, Paris, Louvre

477. VIRGILIO ROMANO *House in Vicolo del Campanile*, Rome

478. PERUZZI *Sala delle Prospettive*, Rome, Villa Farnesina

479. PERUZZI   *Sala delle Prospettive*, Rome, Villa Farnesina

480. PERUZZI   *Sala delle Prospettive* (detail)

481. PERUZZI *Death of Adonis*, Sala delle Prospettive

482. PERUZZI *Procession of Bacchus*, Sala delle Prospettive

483. PERUZZI *Ducalion and Pyrrha*, Sala delle Prospettive

485. PERUZZI *Apollo*, Sala delle Prospettive

484. PERUZZI *Venus and Cupid*, Sala delle Prospettive

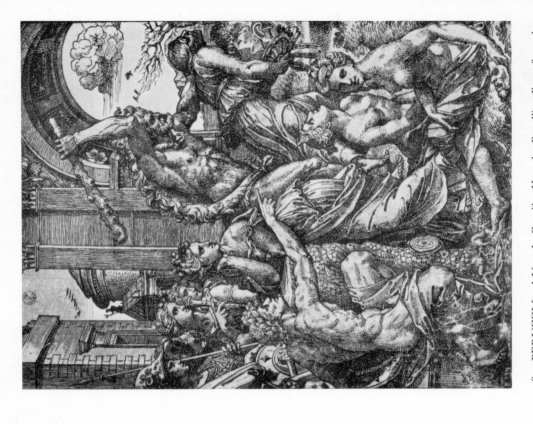

487. PERUZZI [and Ugo da Carpi] *Hercules Expelling Envy from the Temple of the Muses*

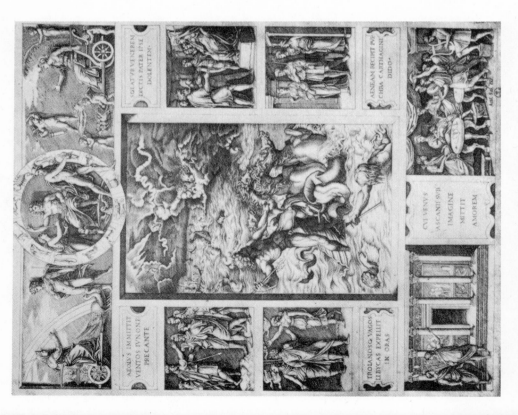

486. PERUZZI (?) [and Marcantonio] *Quos Ego*

488. PERUZZI   *Apollo and the Muses*, Florence, Pitti

489. PERUZZI *Country Festival*, Florence, Uffizi

490. PERUZZI *Ponzetti Chapel* (vault frescoes), Rome, S.M. della Pace

491. PERUZZI *Ponzetti Chapel* (vault frescoes, detail)

492. PERUZZI *Ponzetti Chapel* (altar fresco)

493. PERUZZI *Portrait of a Carmelite*, Rome, Art Market

494. PERUZZI *Presentation of the Virgin*, Rome, S.M. della Pace

495. PERUZZI [and assistants] *Decoration in the Palazzo della Cancelleria, Rome*

496. PERUZZI [and assistants]   Decoration in the Palazzo della Cancelleria, Rome

497. PERUZZI *Joseph put into the Well*, Cancelleria

498. PERUZZI *Meeting of Solomon and Sheba*, Cancelleria

499. PENNI (?) *Separation of Light and Darkness*, Rome, Vatican, Logge

500. GIULIO *Expulsion*, Logge

501. GIULIO *God Appearing to Isaac*, Logge

502. GIULIO *Isaac and Rebecca*, Logge

503. GIULIO [with Polidoro ?] *Jacob and Rachel*, Logge

504. GIULIO (?) [with Perino] *Flight of Jacob*, Logge

505. GIULIO (?) *Crossing of the Red Sea*, Logge

506. GIULIO (?) [with Polidoro] *Moses Striking Water from the Rock*, Logge

507. GIULIO *Moses Receiving the Tablets of the Law*, Logge

508. GIULIO (?) *Adoration of the Golden Calf*, Logge

509. PERINO *Fall of Jericho*, Logge

510. PERINO *Joshua Stays the Sun*, Logge

511. PERINO *Division of the Lands*, Logge

512. PERINO *David and Goliath*, Logge

513. PERINO *David and Bathsheba*, Logge

514. PELLEGRINO (?) *Judgment of Solomon*, Logge

515. POLIDORO   *Meeting of Solomon and Sheba*, Logge

516. POLIDORO   *Building of the Temple*, Logge

517. PERINO *Adoration of the Shepherds*, Logge

518. PERINO *Adoration of the Kings*, Logge

519. PERINO *Baptism of Christ*, Logge

520. PERINO *Last Supper*, Logge

521. PENNI *Study for Separation of Light and Darkness*, London, British Museum

522. PENNI *Study for Expulsion*, Windsor, Royal Library

523. PENNI *Study for Jacob's Dream*, London, British Museum

524. PENNI *Study for Adoration of the Golden Calf*, Florence, Uffizi

525. PERINO *Study for Division of the Lands*, Windsor, Royal Library

526. PERINO *Lamentation over Christ*, Paris, Louvre
RAPHAEL SCHOOL

527. PERINO *Pietà*, Rome, Sto. Stefano del Cacco

528. PERINO   *Adoration of the Child*, Rome, Borghese 464

VI
CLIMAX AND CRISIS IN FLORENCE AND THE GENERATION OF FLORENTINE MANNERISM (c. 1514-1520)

529. FRA BARTOLOMMEO *Virgin and Child*, Florence, Convent of S. Marco

531. FRA BARTOLOMMEO Study for the Madonna della Misericordia, Florence, Uffizi

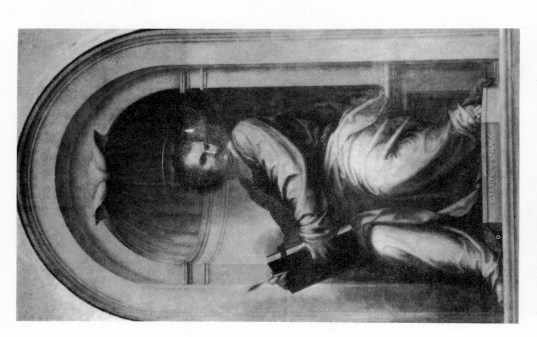

530. FRA BARTOLOMMEO St. Mark Evangelist, Florence, Pitti

532. FRA BARTOLOMMEO *Madonna della Misericordia*, Lucca, Pinacoteca

533. FRA BARTOLOMMEO *Annunciation*, Le Caldine, Convento della Maddalena

534. FRA BARTOLOMMEO *Study for the Caldine Annunciation*, Florence, Uffizi

535. FRA BARTOLOMMEO *Annunciation Altar*, Paris, Louvre

536. FRA BARTOLOMMEO *Salvator Mundi*, Florence, Pitti

538. FRA BARTOLOMMEO *Study for Salvator Mundi*, Amsterdam, Rijksmuseum

537. FRA BARTOLOMMEO *Studies for the Salvator Mundi and Other Projects*, Florence, Uffizi

539. FRA BARTOLOMMEO *Presentation in the Temple*, Vienna, Kunsthistorisches Museum

540. FRA BARTOLOMMEO *Isaiah*, Florence, Academy

541. FRA BARTOLOMMEO *Job*, Florence, Academy

543. FRA BARTOLOMMEO *Study for the Assumption*, Munich, Graphische Sammlung

542. FRA BARTOLOMMEO [with Fra Paolino] *Assumption of the Virgin*, Naples, Galleria Nazionale

545. FRA BARTOLOMMEO *Madonna and Child with Elizabeth and John*, London, Kenwood, Iveagh Bequest

544. FRA BARTOLOMMEO *Holy Family*, Rome, Galleria Nazionale

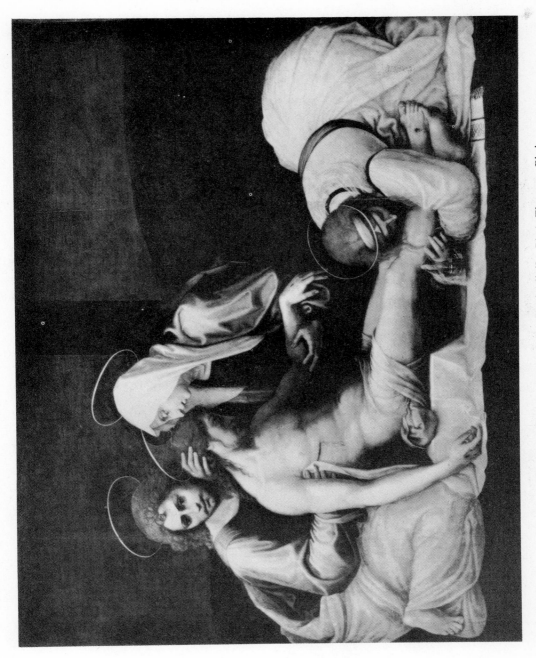

546. FRA BARTOLOMMEO [with Bugiardini] *Pietà*, Florence, Pitti

547. FRA BARTOLOMMEO *Noli Me Tangere*, Le Caldine, Convento della Maddalena

549. FRA PAOLINO *Holy Family with St. John and Angels*, Rome, Galleria Doria

548. FRA PAOLINO *Crucifixion*, Siena, Sto. Spirito

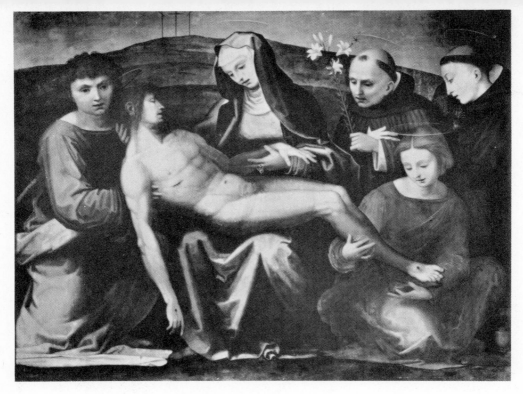

550. FRA PAOLINO *Pietà*, Florence, Museo di S. Marco

551. ANDREA DEL SARTO *Scenes from the Life of St. John Baptist*, Florence, Scalzo

552. ANDREA DEL SARTO *Preaching of St. John Baptist*, Florence, Scalzo

553. ANDREA DEL SARTO *Justice*, Florence, Scalzo

554. ANDREA DEL SARTO *Charity*, Florence, Scalzo

555. ANDREA DEL SARTO *Chiaroscuro Panel* (from the Leo X Festival Decoration ?), Florence, Uffizi

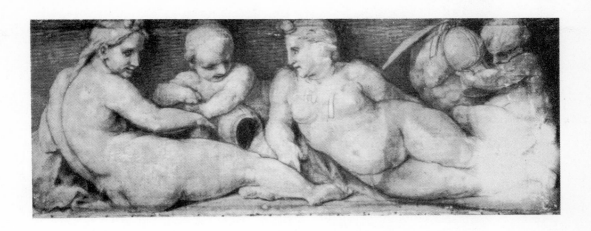

556, 557. ANDREA DEL SARTO *Chiaroscuro Panels* (from the Leo X Festival Decoration ?), Florence, Uffizi

558. ANDREA DEL SARTO *Holy Family with St. Catherine*, Leningrad, Hermitage

560. ANDREA DEL SARTO *Study for the Pietà*, Florence, Uffizi

559. A. VENEZIANO [after Andrea del Sarto] *Pietà*

561. ANDREA DEL SARTO *Redeemer*, Florence, SS. Annunziata

562. ANDREA DEL SARTO *Holy Family*, Paris, Louvre 1515

563. ANDREA DEL SARTO *Baptism of the Multitude*, Florence, Scalzo

564. ANDREA DEL SARTO *Capture of St. John*, Florence, Scalzo

565. ANDREA DEL SARTO *Madonna of the Harpies*, Florence, Uffizi

566. ANDREA DEL SARTO *Disputation on the Trinity*, Florence, Pitti

567. ANDREA DEL SARTO
Portrait of a Sculptor (?),
London, National Gallery

568. ANDREA DEL SARTO
Study for Portrait of a Sculptor, Florence, Uffizi

569. ANDREA DEL SARTO *Caritas*, Paris, Louvre

570. ANDREA DEL SARTO *Madonna and Child with St. John*, Rome, Borghese 334

571. ANDREA DEL SARTO *Madonna with St. John and Angels* (?), London, Wallace Collection

572. ANDREA DEL SARTO *Pietà*, Vienna, Kunsthistorisches Museum

573. ANDREA DEL SARTO *Tribute to Caesar* (in its original dimensions), Poggio a Cajano

574. FRANCIABIGIO *Annunciation*, Turin, Galleria Sabauda

575. FRANCIABIGIO *Angel*, Florence, Sto. Spirito

576. FRANCIABIGIO *St. Job Altar*, Florence, Uffizi

577. FRANCIABIGIO *Meeting of Christ and St. John Baptist*, Florence, Scalzo

578. FRANCIABIGIO(?) *Madonna and Child with St. John*, Vienna, Kunsthistorisches Museum 208

579. FRANCIABIGIO *Benediction of St. John by Zachary*, Florence, Scalzo

580. FRANCIABIGIO
Leda, Brussels, Museum

581. FRANCIABIGIO *Madonna*, Bologna, Pinacoteca 294

582. FRANCIABIGIO *Triumph of Caesar* (in its original dimensions), Poggio a Cajano

583. FRANCIABIGIO *Self-Portrait*, New York, Hunter College, ex Kress Collection

584. FRANCIABIGIO *Portrait of Caradosso*, London, Art Market

586. FRANCIABIGIO *Portrait of a Fattore*, Hampton Court

585. FRANCIABIGIO(?) *Portrait of a Man*, Vienna, Liechtenstein Collection

587a. BUGIARDINI *Temptation in the Garden of Eden*, New York, Private Collection

588. BUGIARDINI
St. Sebastian, New York,
Kress Collection

587b. BUGIARDINI *Temptation in the Garden of Eden*, New York, Private Collection

589. BUGIARDINI
*Madonna and Child with the
Infant St. John*, Dunblane,
Stirling Collection

590. BUGIARDINI *Madonna and Child with the Infant St. John*, Florence, Uffizi

591. BUGIARDINI *Madonna and Child with the Infant St. John*, Allentown (Pennsylvania) Museum, Kress Collection

592. RIDOLFO GHIRLANDAIO *Coronation of the Virgin*, Florence, S.M. Novella, Cappella del Papa

593. RIDOLFO GHIRLANDAIO *Resuscitation of a Youth by St. Zenobius*, Florence, Academy

594. RIDOLFO GHIRLANDAIO *Translation of the Body of St. Zenobius*, Florence, Academy

595. RIDOLFO GHIRLANDAIO *Madonna with Six Saints*, Pistoia, Museum

597. RIDOLFO GHIRLANDAIO *Portrait of a Man*, Florence, Galleria Corsini

596. RIDOLFO GHIRLANDAIO *Girolamo Benivieni* (?), London, National Gallery

598. RIDOLFO GHIRLANDAIO *Portrait of a Man*, Florence, Torrigiani Collection

599. RIDOLFO GHIRLANDAIO *Pietà*, Colle di Val d'Elsa, S. Agostino

600. GRANACCI *Madonna with Sts. Francis and Zenobius*, Florence, Academy

601. GRANACCI *Holy Family*, Boughton House, Duke of Buccleuch

602. GRANACCI *Madonna and Child*, San Francisco, Palace of the Legion of Honor

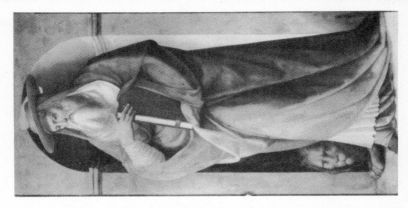

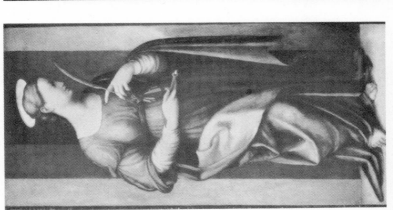
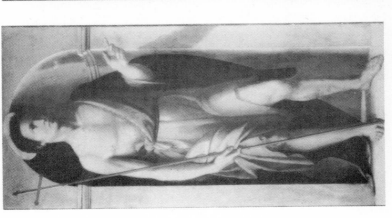

603. GRANACCI Sts. John, Apollonia, Mary Magdalen, and Jerome (from the *Apollonia Altar*), Munich, Pinakothek

604. GRANACCI *Predella Panel* (from the *Appollonia Altar*), Florence, Academy

605. GRANACCI *Predella Panel* (from the *Apollonia Altar*), Florence, Academy

606. GRANACCI *Predella Panel* (from the *Apollonia Altar*), Florence, Academy

607. GRANACCI *Entry of Charles VIII into Florence*, Florence, Museo Mediceo

608. GRANACCI *Joseph Presents His Father to Pharaoh*, Florence, Uffizi

609. GRANACCI *Arrest of Joseph*, Florence, Uffizi

610. GRANACCI *Madonna with St. John*, (formerly) Munich, A. S. Drey
MASTER OF THE KRESS LANDSCAPES

611. GRANACCI *Madonna with Four Saints*, Montemurlo, Pieve

612. SOGLIANI *Madonna with St. John*, Baltimore, Walters Art Gallery

613. SOGLIANI *Madonna with St. John*, Brussels, Museum

614. SOGLIANI *Madonna with St. John*, Turin, Galleria Sabauda

615. SOGLIANI *S. Acasio Altar*, Florence, S. Lorenzo

616. SOGLIANI *S. Brigitta Altar*, Florence, Academy

617. PULIGO [on Sarto's cartoon] *Holy Family*, London, National Gallery

618. ANDREA [with Puligo] *Story of Joseph* (1), Florence, Pitti 87
ANDREA DEL SARTO

619. ANDREA [with Puligo] Story of Joseph (2), Florence, Pitti 88
ANDREA DEL SARTO

621. PULIGO *Madonna with St. John and Angels*, Florence, Galleria Corsini

620. PULIGO (?) *Madonna with St. John*, Florence, Pitti 242

622. PULIGO *Adoration of the Kings* (formerly) Milan, Crespi Collection
MASTER OF THE KRESS LANDSCAPES (?)

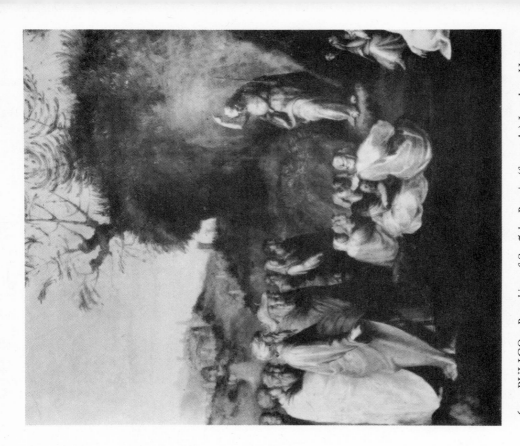

624. PULIGO *Preaching of St. John Baptist* (formerly) London, Henry Harris Collection
ANTONIO DI DONNINO

623. PULIGO *Deposition*, Venice, Seminario
ANTONIO DI DONNINO

625. PULIGO *History of Joseph*, Rome, Borghese 463
MASTER OF THE KRESS LANDSCAPES

626. PULIGO *Apollo and Daphne*, Florence, Galleria Corsini
ANTONIO DI DONNINO

627. BACCHIACCA *Deposition*, Bassano, Museo Civico

628. BACCHIACCA *Adam and Eve*, Philadelphia Museum, Johnson Collection

629. BACCHIACCA *Story of Joseph*, London, National Gallery 1218

630. BACCHIACCA *Story of Joseph*, Rome, Borghese

631. BACCHIACCA *Story of Joseph*, London, National Gallery 1219

632. BACCHIACCA *Story of Joseph*, Rome, Borghese

633. BACCHIACCA *Deposition*, Florence, Uffizi

635. BACCHIACCA *Leda*, Rotterdam, Boymans Museum, van Beuningen Collection

634. BACCHIACCA *Creation of Eve*, Stockholm, Private Collection

636, 637. BACCHIACCA *Predella Panels* (from the *S. Acasio Altar*), Florence, Uffizi

638. BACCHIACCA *Legend of the Dead King*, Dresden, Gallery

639. PONTORMO *Visitation*, Florence, SS. Annunziata

640. PONTORMO *Visitation* (detail)

641. PONTORMO *Cappella del Papa* (vault), Florence, S.M. Novella

642. PONTORMO *St. Veronica*, Florence, S.M. Novella, Cappella del Papa

643. PONTORMO *Joseph Revealing Himself to His Brothers*, Henfield, Lady Salmond

645. PONTORMO *The Butler Restored and the Baker Led to Execution,* Henfield, Lady Salmond

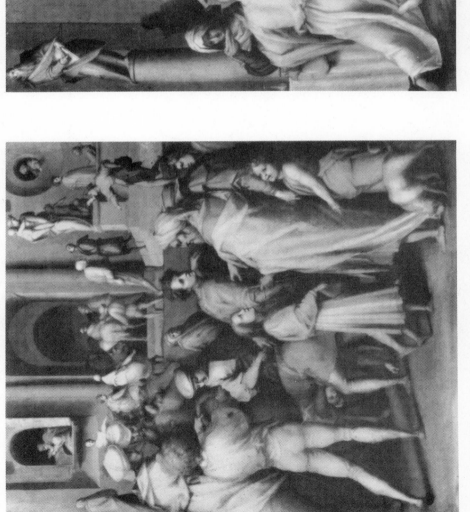

644. PONTORMO *Joseph Sold to Potiphar,* Henfield, Lady Salmond

646. PONTORMO *Madonna and Saints*, Florence, S. Michele Visdomini

647. PONTORMO *Pietà* (predella for the Visdomini Altar), Dublin, National Gallery
from designs by PONTORMO

648, 649. PONTORMO *St. Lawrence; St. Francis* (portions of predella), Dublin, National Gallery
from designs by PONTORMO

650. PONTORMO *Joseph in Egypt*, London, National Gallery

651. PONTORMO *Cosimo de' Medici*, Florence, Uffizi

653. PONTORMO *Portrait of a Musician (Francesco dell' Ajolle?)* Florence, Uffizi

652. PONTORMO *Study for a Portrait of Piero de' Medici,* Rome, Galleria Corsini

654. PONTORMO *Study for St. John Evangelist*, Florence, Uffizi

655. PONTORMO *St. John Evangelist*, Empoli, Collegiata

656. PONTORMO *St. Michael*, Empoli, Collegiata

657. PONTORMO *Study for a Pietà*, Florence, Uffizi

658. PONTORMO *St. Anthony Abbot*, Florence, Uffizi

660. BERRUGUETE *Madonna*, Florence, Uffizi

659. BERRUGUETE *Salome*, Florence, Uffizi

661. BERRUGUETE *Madonna with St. John*, Florence, Palazzo Vecchio, Loeser Collection

662. ROSSO *Madonna in Glory*, Leningrad, Hermitage

663. ROSSO *Portrait of a Young Man*, Berlin, Museums

664. ROSSO *Assumption*, Florence, SS. Annunziata

665. ROSSO *Assumption* (detail)

666. ROSSO *Memento Mori*, Florence, Uffizi

667. ROSSO *Madonna and Saints (S.M.Nuova Altar)*, Florence, Uffizi

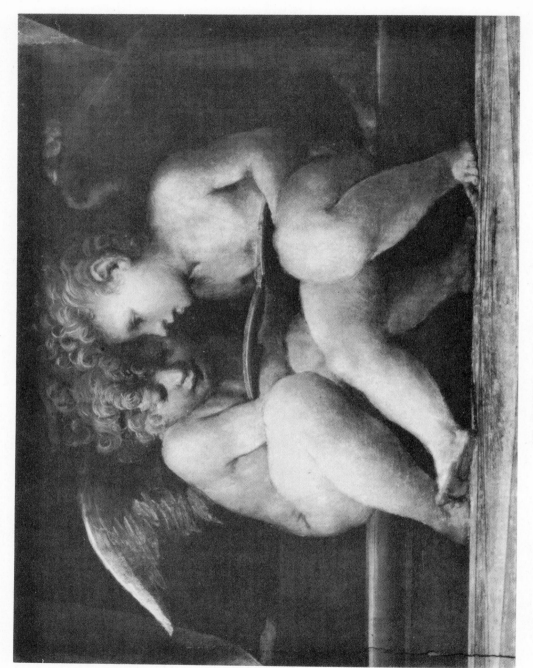

668. ROSSO *S.M. Nuova Altar* (detail)

VII

EPILOGUE: THE ASCENDANCY OF MANNERISM

(SOME EVENTS OF 1521)

669. ROSSO *Madonna with Sts. John Baptist and Bartholomew*, Villamagna (near Volterra), Pieve

670. ROSSO *Deposition*, Volterra, Museum

671. ROSSO *Deposition* (detail)

672, 673. ROSSO *Deposition* (details)

674. PONTORMO Design for Wall at Poggio a Cajano, London, British Museum

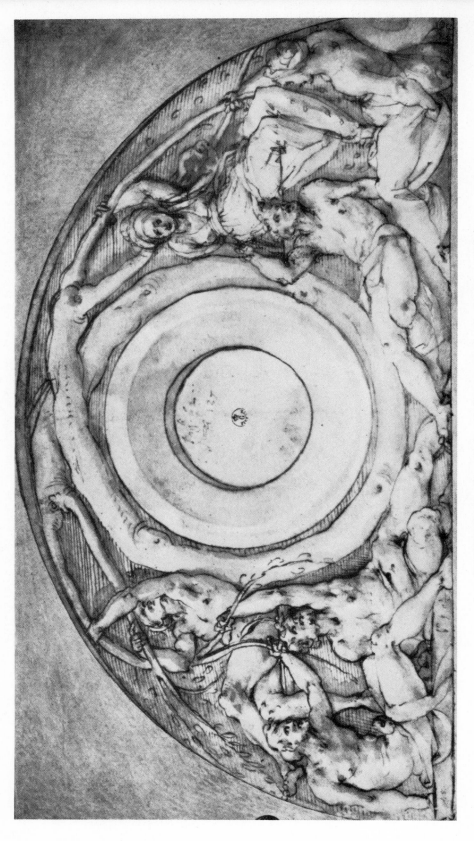

675. PONTORMO *First Project for Lunette at Poggio*, Florence, Uffizi

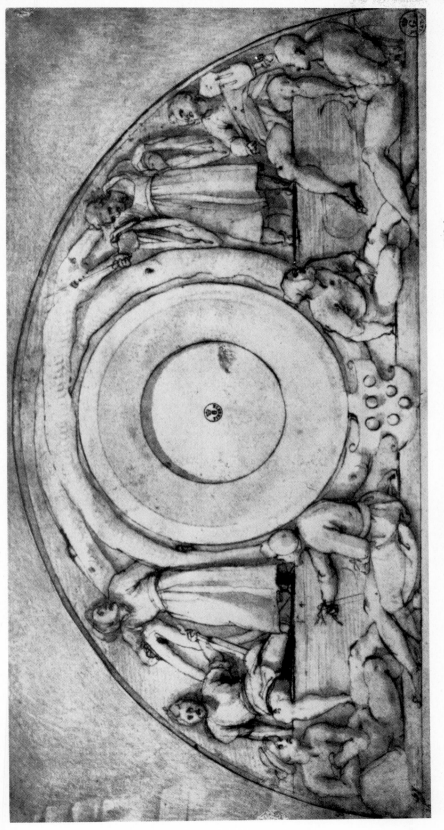

676. PONTORMO Second Project for Lunette at Poggio, Florence, Uffizi

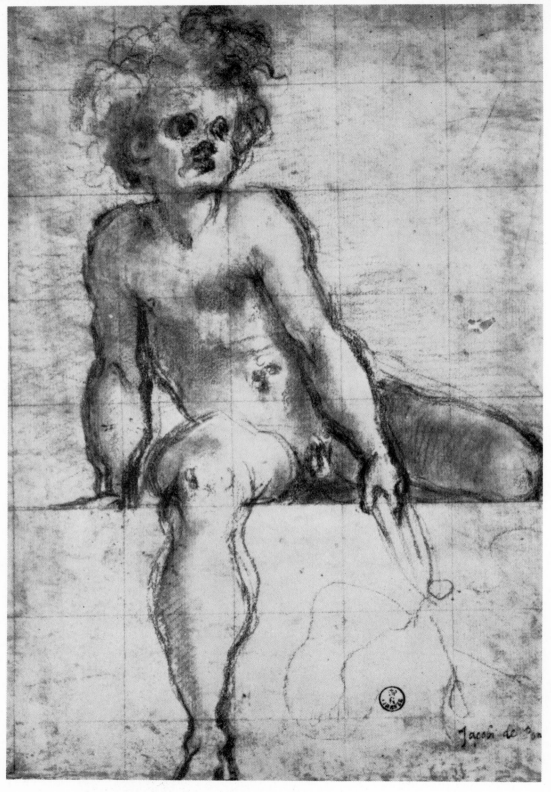

677. PONTORMO *Study for Lunette at Poggio*, Florence, Uffizi

678. PONTORMO *Vertumnus and Pomona*, Poggio a Cajano

679a. PONTORMO *Vertumnus and Pomona*, Poggio a Cajano

679b. PONTORMO *Vertumnus and Pomona*, Poggio a Cajano

680. PONTORMO *Vertumnus and Pomona* (detail)

681. GIOVANNI DA UDINE [and assistants] *Loggia*, Rome, Villa Madama

682. GIOVANNI DA UDINE [with Peruzzi] *Loggia* (central vault), Villa Madama

683, 684. GIOVANNI DA UDINE [with Peruzzi] *Loggia* (side vaults), Villa Madama

685, 686. PERUZZI *Loggia* (details), Rome, Villa Madama

687. GIOVANNI DA UDINE [with Perino] *Sala dei Pontefici*, Rome, Vatican

688. GIOVANNI DA UDINE [with Perino] *Sala dei Pontefici* (detail)

689, 690. PERINO *Sala dei Pontefici* (details)

691. PERINO *Sala dei Pontefici* (central panel of the vault)

692. GIULIO *Justice* (trial figure), Rome, Vatican, Sala di Costantino

693. GIULIO [and assistants] *Sala di Costantino*, Rome, Vatican

694. GIULIO [and assistants] *Sala di Costantino*, Rome, Vatican

695. GIULIO *Allocutio*, Sala di Costantino

696. GIULIO [and assistants] *The Battle of Constantine*, Sala di Costantino

697. GIULIO *St. Peter with Ecclesia and Aeternitas*, Sala di Costantino

698. GIULIO *Leo X as Clement I with Moderatio and Comitas*, Sala di Costantino

699. GIULIO [and assistants]   *Battle of Constantine* (detail), Sala di Costantino

700. POLIDORO   *Basamento*, Sala di Costantino

72 73 74   12 11 10 9 8 7 6 5 4 3 2 1